ALIEN VOLCANOES

© Michael Carroll 07

ALIEN VOLCANOES

Printed in China on acid-free paper
9 8 7 6 5 4 3 2 1

The Johns Hopkins University Press
2715 North Charles Street
Baltimore, Maryland 21218-4363
www.press.jhu.edu

Library of Congress Cataloging-in-Publication Data
Lopes, Rosaly M. C., 1957–
Alien volcanoes / Rosaly M. C. Lopes and Michael W. Carroll; with illustrations by Michael W. Carroll; foreword by Arthur C. Clarke.
p. cm.
Includes index.
ISBN-13: 978-0-8018-8673-7 (hardcover : alk. paper)
ISBN-10: 0-8018-8673-2 (hardcover : alk. paper)
1. Volcanoes. 2. Planetary volcanoes. 3. Solar system. I. Carroll, Michael W., 1955– II. Title.
QE522.L67 2008
551.210999'2—dc22 2007031883

A catalog record for this book is available from the British Library.

Frontispiece: Cryovolcanoes erupt on Saturn's ice moon Enceladus.

Special discounts are available for bulk purchases of this book. For more information, please contact Special Sales at 410-516-6936 or specialsales@press.jhu.edu.

The Johns Hopkins University Press uses environmentally friendly book materials, including recycled text paper that is composed of at least 30 percent post-consumer waste, whenever possible. All of our book papers are acid free, and our jackets and covers are printed on paper with recycled content.

Page 152 constitutes an extension of this copyright page.

CONTENTS

FOREWORD

I have always been interested in volcanoes. And over the years I have had the opportunity to visit a few well-behaved ones, especially in Hawaii. My 1953 novel *Childhood's End* opens with the imminent launch of a spaceship on a volcanic island in the middle of the Pacific.

Thanks to the robot space probes that have explored most of our solar system, we now know that our volcanoes have counterparts on other planets and moons—some of them large enough to dwarf any on earth. *Alien Volcanoes* offers a closer look at these remarkable extraterrestrial geological features. It is a collaboration between planetary scientist Rosaly Lopes and space artist–writer Michael W. Carroll.

Until a generation ago, the only volcanoes we knew were on our home planet. But in less than four decades we have discovered volcanic features—and sometimes, volcanic activity—on several planetary bodies.

Mars was the first planet to be slowly unveiled by our space probes, and I can still remember our initial disappointment. Three U.S. spacecraft—Mariners 4, 6, and 7—reached Mars during the 1960s. They radioed back images of a flat landscape peppered with impact craters, almost indistinguishable from the lunar surface. Between them, the three Mariners returned about two hundred images, covering a mere 10 percent of the planet. They turned out to be unexciting and unremarkable—and we expected more of the same dreary, flat terrain to emerge from Mariner 9.

On November 12, 1971, the Jet Propulsion Laboratory (JPL) brought together four incurable Mars addicts: Carl

Sagan, Bruce Murray (who later became JPL's director), Ray Bradbury, and me. Under the chairmanship of the *New York Times'* distinguished science writer Walter Sullivan, we conducted a free-ranging discussion before a large and enthusiastic audience. The timing was impeccable. Next day, Mariner 9 was due to go into orbit around Mars and conduct the first detailed reconnaissance of the planet.

Ray and I talked about how science fiction had inspired the scientific exploration of Mars—and speculated how new scientific findings could inspire a whole new generation of science fiction writers and artists. (As indeed happened.) All of us knew that November evening how important the mission might be, but I doubt if any would have dared to predict the full extent of its success. True, there were no irate Martians carrying banners "Earthlings go home!" But what we saw was exciting enough. Mariner 9 sent back a total of seven thousand images over several months—revealing one of the most spectacular landscapes in the solar system.

While I fully shared the excitement of these new findings, the unveiling of Mars caused me some embarrassment as a writer. My first full-length novel, *The Sands of Mars* (1951), contained the sentence, "There are no mountains on Mars." Well, the mightiest of all the Martian volcanoes, Olympus Mons, is almost three times the height of Everest and about 600 km (370 mi) across—it is also the largest known volcano in the solar system. In subsequent editions, I apologized to readers for not having anticipated this Martian detail.

Nineteenth-century astronomers, peering at the tiny and often blurred image of Mars, noted a bright spot to which they gave the astonishingly prescient name Nix Olympica, or the Snows of Olympus.

Now that the real nature of this marking is known, the name has been changed to Olympus Mons—Mount Olympus. Though most of its brilliance is due to cloud cover, there is probably snow on that awesome summit—albeit snow of carbon dioxide, not water.

But Olympus Mons is by no means the only volcano among the planets and moons of our cosmic neighborhood. Since 1971 a succession of planetary probes has brought us images from other planets and their many moons. Ninety percent of the earth's twin, Venus, is covered by volcanoes, and volcanic domes, cones, and mountains have been found on the earth's moon—and probably exist on Mercury. There are also active volcanoes on Jupiter's moon Io.

Despite the remarkable accomplishments of our exploratory space missions, we still have only distant views of the most intriguing of our solar system's volcanoes. The scientist can tell us how these wonders might appear up close and describe probable causes and effects; the artist can take us to these distant sites, put us on the surface, and show us these alien volcanoes on a human scale. *Alien Volcanoes* treats us to a powerful, elegant combination of words and pictures, art and science.

Beyond the earthlike worlds of the inner solar system, we find volcanoes of a different sort. These exploding mountains are energized by cryogenic lavas of water, nitrogen, methane, and ammonia. Out in the dark, frozen reaches of space, other volcanoes await future space explorers.

Fasten your seatbelts!

Arthur C. Clarke

ACKNOWLEDGMENTS

Our thanks to the many researchers who reviewed various portions of the text, including, Claudia Alexander, Jayne Aubele, Robert Carlson, Larry Crumpler, Kevin Hand, Ralph Lorenz, and Paul Weissman.

INTRODUCTION

Half a century ago, at the dawn of space exploration, a handful of planetary scientists pondered the idea of volcanoes on other worlds. Through the telescope, our own moon displayed a face blanketed by what seemed to be extensive lava flows. Some craters perched atop mounds looking like terrestrial cinder cones (though later most of them were shown to be impact craters). Other pits looked like they were chained together, much like a collapsed lava tube. There were even reports of transient lunar phenomena, strange glowing lights emanating from shadowed crater floors and canyon walls. These were never verified, but remain tantalizing. Many years later, Apollo 15 astronauts landed by Hadley Rille, one of the collapsed lava tubes. The Apollo missions enhanced our understanding of the geology of the moon, including the role of volcanism in creating the moon's surface as we see it today.

If the earth's moon had been imbued with exploding mountains and rushing lava flows, what of other worlds? Did Venus have her own Vesuvius, glowering beneath her clouds? Early radar studies of Venus revealed large mountains, but the origin of those objects was the topic of heated debate. Did nearby Mars sport the Martian equivalent of Fujiyama? The first three successful Mars probes shed no light on the subject as they imaged only part of the surface.

As space exploration increased our knowledge of other planets and their moons, we discovered volcanoes on many alien worlds. In 1971 the Mars orbiter Mariner 9 revealed Everest-tall volcanoes—the largest yet found in the solar system—rearing out of a planetwide dust storm. Closer to home, spacecraft circled and even landed on Venus, sending back data that suggested the presence of volcanic rocks. In the early 1990s the Magellan spacecraft, carrying cloud-piercing radar, revealed that the surface of Venus was covered with volcanic features, some of which appear to be young in geologic terms. In fact, more than 90 percent of the Venusian surface has been transformed by volcanic eruptions.

Spacecraft ventured beyond the asteroids, into the outer reaches of our planetary system. Voyager and Galileo revealed Jupiter's moon Io to be the quintessential site of alien volcanoes, with geyserlike plumes erupting 500 km (300 mi) into the airless sky. In the frigid outer solar system, other moons showed signs of past eruptions of liquid water. On the distant moon Triton, geysers spouted gaseous nitrogen. Geysers of water vapor have been found on Saturn's moon Enceladus by the Cassini spacecraft. Everywhere we look, we see evidence of ancient or active eruptions of molten rock, water, and even more exotic lavas.

The volcanoes closest to home are the most familiar in nature. Mercury, Venus, Earth, and Mars all have volcanoes driven by the heat of radioactivity deep inside them. Heat from the initial formation of these worlds, coupled with the simmering of radioactive minerals within, is a recipe for violent venting of internal heat in the form of erupting mountains and fountaining streams of liquid rock. The volcanoes of the outer solar system are far more alien than those closer to home. To understand and appreciate the complexities of those alien volcanoes, we must first look to our home world. Here, volcanoes have played a colorful part in the history of the planet and its inhabitants.

In this book we combine some of the best photos of alien volcanoes with paintings of views not yet seen by human explorers. Often, our images of volcanoes are limited to distant observations, sometimes not even at visible wavelengths. On those occasions where we have gotten close enough to recognize volcanic features such as lava flows and cinder cones, the view is from above. But the artist can imagine what the robotic eye cannot see. And we offer the reader the best of what space exploration has revealed and the artist's skill has engendered.

Biologists have a term called the "edge effect," which refers to the effect caused by the juxtaposition of two disparate environments, for example, where the ocean meets the desert. In a natural environment, life is richer and more varied at the junction of the two biomes. The edge effect can also be applied to art and science. When these two seemingly disparate disciplines meet, beautiful things happen. The paintings in these pages are a result of the edge effect created by the confluence of art and science. Someday, humans will witness firsthand an eruption on Io or a nitrogen geyser on Triton, but for now, we must be content with exploration of these magnificent volcanic worlds through the eyes of the spacecraft, the scientist, and the artist.

ALIEN VOLCANOES

1 VOLCANOES IN HISTORY

Long before there were people to witness them, before oceans washed across the face of our planet Earth, before life took hold here, there were volcanoes. They brought glowing energy and materials from deep within the earth, enriching the environment and sculpting the landscape. Much of the air we breathe today comes from the atmospheric building blocks of these early eruptions.

Early in earth's history, high magnesium lava resulted in superheated eruptions as hot as 1800K (1527°C [2780°F]). Despite their earthly origins, these eruptions were quite alien by today's standards. Known as ultrabasic (or ultramafic) eruptions, these volcanoes were as hot as liquid wrought iron. The addition of a constant hail of molten rock in fire fountains and rushing lava waves would have resembled a medieval painting of Hell.

Where did all this heat come from? The answer lies, in large part, with the early formation of our solar system. Just over 4,500 million years ago, a traveler flying by the solar system would have seen a vast cloud of dust and rubble pinwheeling around an infant sun. The cloud was laced with chunks of debris ranging in size from the particles in cigarette smoke to asteroids hundreds of miles across. Grain stuck to grain, and boulder nudged boulder, expanding into piles of rubble. The larger bodies, having substantial gravitational pull, progressively attracted more material around them, growing ever larger and denser. The earth began in this humble fashion. Some hundred thousand years after the birth of the sun, the earth was undifferentiated, meaning that its materials were well mixed. The heavy nickel-

Early in earth's history, the planet cooled and the first oceans began lapping at the shore. On the moon, oceans of molten rock spread across the barren surface of earth's satellite.

FIRE AND BRIMSTONE

Hell was a popular subject in medieval paintings. The Dutch artist Hieronymus Bosch rendered a nightmarish scene of Hell in the right panel of the triptych titled *The Garden of Earthly Delights.* Note the volcanic forms in the background.

(*opposite, top*) Throughout its history, the earth has been peppered by space debris such as meteors, comets, and asteroids. The Barringer Meteor Crater near Winslow, Arizona, is 1,300 m (4,265 ft) in diameter and 174 m (571 ft) deep.

(*opposite, bottom*) Modern lava fields, such as this one on Kilauea, Hawaii, give us a sense of what the surface of the primordial earth was like.

iron core of our present-day earth was still to come. As more material stuck to the protoearth, its gravity increased. The same process was happening elsewhere, and soon there were many orbiting bodies dozens of miles across. With its increasingly attractive personality, the earth began to pull in larger objects at greater velocities.

Early on, these collisions were low-energy meetings concluded—more often than not—by a gentle bump. Then, the rain began to fall. This was no gentle spring shower, but rather a storm of rock and metal. Space debris fell to the surface of the world in massive quantities, with devastating effect. By 30 to 40 million years after the solar system's birth, the earth was the size of Mars. Its gravity no longer gently coaxed passing asteroids to itself. Instead, the encounters became fierce. Infalling asteroids and meteors slammed into the surface, heating it to the melting point. Some impactors were the size of respectable moons and generated tremendous energy upon landfall. This heat helped to keep earth's interior fairly fluid, so that heavier

materials sank—becoming a dense core—in a process called differentiation. It took between 50 and 90 million years for the earth to grow from a grain of sand to its present size. The violence associated with earth's formation was astounding.

The earth was left with a molten surface. Waves of a lava sea sloshed against newly cooled stone. Titanic explosions rocked the fragile crust. Infalling asteroids triggered giant waves of glowing liquid rock. At this time, about 4,500 million years ago, the earth was fully differentiated, with a dense core and less dense outer layers, each having a slightly different chemical makeup. The planet's internal structure was essentially the same as it is today. At its center sat a core of iron and nickel nearly the size of Mars. The core became liquid at its edge where lower pressure allowed the searing nickel-iron to liquefy. Outside the core lay a region called the mantle, dense rock rich in metals and high in temperature. This region would be important to future volcanoes. At the top of the mantle, a crust formed. Today, the crust is many kilometers thick, and is made up of rock that includes granite (less dense rock that makes up much of our continents), basalt (which erupted from volcanoes), and sedimentary (layered) rock. The crust is thickest under the continents, where its 40-km (25-mi) depth upholds continental slabs like ice floating on a lake. Under some mountain ranges the crust may be as deep as 80 km (50 mi). It is this crust that holds in much of the heat from the fires of creation.

But the crust is only part of the heat story. Radioactive elements within our planet contribute their own heat, a variable that was not understood until fairly recently. In 1846, the Scottish physicist Lord Kelvin judged the earth's temperature to be too cool for it to be of any great age. He estimated that a body the size of the earth, gradually cooling from a starting point as hot as the sun, could only be 100 million years old. Kelvin did not know that the outer layers of the earth insulate its interior heat and that internal radioactivity replenishes some of it.

It is a simple physical fact of nature that the vast quantities of heat held within the earth want to get out. Everything in nature seeks balance. It's cold out in space, and it's hot inside the earth. Nature wants to correct this imbalance. Volcanoes are links between the "outside world" and the inferno of the earth's interior mantle. As soon as there was a crust to cap the world's inner heat, there were eruptions through that crust. Volcanoes were born early.

Mercury, Venus, and Mars, known as the terrestrial planets, were formed in much the same way the earth was. The formation of the outer planets was ruled by primordial gases

VOLCANOES AND ICELANDERS

Iceland's starkly beautiful landscape has been shaped by volcanic action. Since the earliest settlement days, Icelanders have harnessed hydrothermal energy from the fissures scattered across the land.

Iceland was settled in the tenth century AD. For many years the Icelanders lived in relative harmony, tolerating a wide spectrum of religious beliefs, from Thor-worship to skeptic to Christian. As Christianity gained in popularity, the question of whether Icelanders should adopt Christianity as the one true religion came to a head in about AD 1000 and the people were to vote on the matter.

Icelanders from all over the island met at the Law Rock (Lögberg) and held a national assembly called the althing. This was democracy in action; discussion took place in a public arena and legal decisions were made by popular vote. A local Icelandic leader was speaking in favor of adopting Christianity when a fountain of fire erupted. What happened next is recorded in the ancient Icelandic text known as Njal's saga:

Then came a man running and saying that earth-fire was coming up in Oelfus and that it would overrun the homestead of Thorod Gothi. Then the heathen men began to say, "It is no wonder that the gods are enraged at such speeches." And Snorri said, "About what were the gods angry when the lava burned the ground on which we now stand?"

and ice that were not present for much of the formation of the inner planets. In the inner solar system, these materials were cleared away by the adolescent sun in an energetic epoch called the T-tauri phase. During this period, increased solar wind pushed light volatiles out into the distant regions of the planetary system, leaving only the heavy materials behind. Those materials formed the building blocks of the rocky inner worlds.

All the inner planets, as well as the earth's moon, appear to have gone through at least early volcanism and magma oceans. Venus and Mars have extensive volcanic structures created over long periods of time. Some of these structures may be dormant or active even today. The dark blotches blanketing our moon are remnants of magma oceans that once swept across its face. At one time in earth's history, elements of volcanism on the glowing surface of the moon could have been observed from the surface of a volcanically active earth. Because the moon was closer to the earth than it is today, its orb would have been an imposing ember in the terrestrial sky.

Volcanoes continued to sculpt and resurface the landscape of our world as the surface cooled, the atmosphere stabilized, and life emerged in the sea and on land. Eventually, humans would arrive to ponder their world and interact with its natural wonders, including volcanoes.

In Iceland, an island nation born from volcanic upheaval, human-volcano interactions have been commonplace. Early Icelandic settlers had a visceral relationship with volcanism. Rich volcanic soil brought life to farmlands. Hot springs saved travelers in the winter and provided places to cook such delicacies as soup made from fjallagras (Icelandic moss) and slatur (sheep leftovers tied up in the stomach and cooked). Lava flows expanded Icelandic borders, forging new lands in the presence of Icelanders. But volcanoes also brought hardship and destruction: farmhouses were lost to lava and ash; eruption-triggered floods and mudslides took many lives. The Laki eruptions of 1783–84 constituted the second largest eruptive event in recorded history. Some 14 cu km (3.4 cu mi) of magma rocketed out of the earth's interior, blanketing 565 sq km (218 sq mi). Although the eruption did not kill anybody directly, about nine thousand people died from the famine that followed this major environmental disaster. Farmlands were destroyed, and most of the livestock on the island perished, poisoned from eating fluorine-contaminated grasses. Icelanders have always had a great respect for and understanding of the volcanoes around them, and today they even harness volcanic power to provide inexpensive heating for their homes.

Volcanoes have always combined beauty and upheaval,

Icelandic hydrothermal sites, or "hot pots," provided welcome warmth for travelers and allowed early settlers to cook food.

as do other major forces of nature. Brazilians refer to this as *belo horrível,* the "beautiful horrible." Pliny the Younger described the eruption of Vesuvius as "devastation of the loveliest of lands," lands made lovely—he might have pointed out—by enriched soil from the very volcano that would destroy them. Pliny's account of the death of his uncle Pliny the Elder during a rescue attempt of the people near Vesuvius in AD 79, is the earliest surviving scientific account of a volcanic eruption. The detailed description actually comes to us through the Roman historian Tacitus, who wrote to Pliny the Elder's eighteen-year-old nephew for details of his friend's death.

Pliny the Elder was overseer of the Roman naval fleet at Misenum, across the Bay of Naples from Mount Vesuvius. On August 24 at 1:00 p.m., an enormous cloud boiled up into the skies over the majestic mountain, looking like a flat-topped Mediterranean pine. As Pliny the Younger wrote, "It rose into the sky on a very long 'trunk' from which spread some 'branches.'... Some of the cloud was white, in other parts there were dark patches of dirt and ash."

Pliny the Elder, who had written a treatise on natural history, was fascinated by the phenomenon and prepared a fast boat to cross the bay for a closer inspection. Fortunately for historians of volcanology, his nephew turned down the invitation to accompany his uncle. Before Pliny the Elder had departed, an urgent message arrived from Rectina, a noblewoman whose villa was somewhere near the volcano. Pliny immediately changed his purpose: his information-gathering plan became a rescue mission. Mobilizing the

Vesuvius erupted in AD 79, engulfing the towns of Pompeii and Herculaneum. The sites, which have been gradually uncovered since the eighteenth century, offer remarkable insight into Roman life in the first century AD.

warships, he sailed across the bay. As the fleet approached within six miles of the coast, searing ash and a hail of pumice bombarded it. The waters near the coast were choked with floating pumice, making landing impossible. Pliny took the fleet to a villa at Stabiae, ten miles south of Vesuvius. A skilled observer, Pliny continuously dictated his observations to a scribe. The situation worsened throughout the night, with sheets of fire visible on the flanks of the great mountain. In the morning, Pliny's party made their way to the harbor. "They tied pillows on top of their heads as protection against the shower of rock. It was daylight now elsewhere in the world, but there the darkness was darker and thicker than any night. But they had torches and other lights." Pliny describes a terrifying, dark cloud "shot through by the curving trajectories of twisting incandescent vapors." Pliny the Elder died in the ashfall, possibly due to complications of asthma or a heart condition. He was one of more than two thousand casualties in the eruption of Vesuvius, whose ash preserved the Roman cities of Pompeii and Herculaneum for future archeologists. Beauty from destruction. *Belo horrível.*

Pliny the Younger was not the first to attempt a scientific discussion of volcanoes. That distinction may fall to Plato (d. 347 BC), who asserted that earthquakes were the result of hot winds imprisoned in underground caverns. These winds were ignited, he suggested, by the burning river Pyriphlegethon and blew from the earth as volcanoes.

A near-contemporary of Plato's was Empedocles of Acagras (492–432 BC). Empedocles died while studying the

active volcano Mount Etna (some say he threw himself into the crater). Half a millennium later, the Greek traveler Strabo described volcanoes as the earth's "safety valves," an advanced idea for its time.

The terrifying destruction of Pompeii is depicted in this painting, *The End of Pompeii.*

The Renaissance saw renewed interest in all things scientific, including the workings of volcanoes. Various proposals eventually led to the idea of Neptunism. This theory suggested that sulfur, pyrites, and other minerals combusted when exposed to underground water and air. Thus, it was supposed that volcanoes were the result of processes near the surface, where air and moisture ignited flammable subterranean matter. As the French naturalist Georges-Louis Leclerc de Buffon wrote in the eighteenth century, "A volcano is an immense cannon whose barrel may be a mile in diameter . . . caused by the fact that in the burning mountains are accumulations of . . . flammable matter which ferment when exposed to air and dampness."

The British scientist James Hutton advocated an alternate theory called Plutonism (after Pluto, god of Hades).

(*top*) Smoke billows from Krakatau, which lies in the Sunda Strait between Java and Sumatra.
(*bottom*) Despite their heat and mass, lava flows like this one at Hawaii's Kilauea are not as dangerous as the rapidly moving nuée ardentes that have plagued sites such as Martinique.

Plutonism asserted that the earth's interior is intensely hot and that this heat escapes through fractures in the ground, releasing molten rock and gas. Many place Hutton's theory as the beginning of modern volcanology. Indeed, Hutton said the earth was like a living machine, powerful and very ancient. His work foreshadowed the groundbreaking theories of Alfred Wegener, two centuries later.

While people were trying to figure out what made volcanoes tick, volcanoes continued to do what volcanoes do best. One of the most famous eruptions in recorded history is that of Krakatau (sometimes mistakenly called Krakatoa), an old volcanic cone on the island of Rakata in Indonesia. Volcanic events had occurred at the site since at least AD 416, when the majority of the island collapsed into the ocean as a 6.4-km-wide (4-mi) submarine crater, or caldera. Rakata rebuilt itself as a merged chain of three volcanoes. The island had been quiet for some two centuries when it became restless again in early 1883. In that year, a series of earthquakes struck the area.

On May 20, 1883, the main volcano erupted. Early that morning, the captain of the German warship *Elizabeth* reported a cloud towering above the uninhabited island. Explosions of searing steam and ash rose 11 km (7 mi) into the sky. Over the next three months, the volcano exploded with increasing frequency and ferocity. Finally, on the afternoon of August 26, 1883, a titanic explosion rocked the island, casting a black cloud 27 km (17 mi) into the skies of Indonesia. Three more explosions occurred the next morning. The last was the worst. The eruption vaporized the northern two-thirds of the island. Great undersea structures of ancient volcanic calderas collapsed, sending a 40-m (131-ft) tsunami up the strait of Sunda. The towering waves killed thirty-six thousand people in Java and the Sumatra islands. The explosion itself was heard three thousand miles away, and its ash was seen on the other side of the globe in New York City. Its explosive force was equivalent to 150 megatons, or 750 Hiroshima atomic bombs. It was the most destructive volcanic event in recorded history.

Krakatau's effects were long lived. Twenty-one cu km (5 cu mi) of the island were powdered and thrown into the upper atmosphere. Shock waves rippled through the atmosphere seven times over the next five days. Within thirteen days, a cloud of dust had formed a band straddling the entire equator. Strange sunset effects were reported for nearly three years. Global temperatures dropped by as much as 1.2°C (2.1°F) and did not return to normal for five years.

Two decades after the explosion of Krakatau, another eruption grabbed the worldwide headlines, mainly because of its swift destruction of a modern city and nearly all its inhabitants. The picturesque island of Martinique, which has been called a bit of France in the Caribbean, has been home to famous artists, including the painter Paul Gauguin and the musician Alexandre Stellio. It was the birthplace of Josephine, wife of Napoleon Bonaparte. At the start of the twentieth century, according to the local newspaper, Martinique's capital, Saint Pierre, had the distinction of being "the first rum exporter in the world."

Early in 1902, fumaroles began to open on the side of

Martinique's Mount Pelée seems placid today, though it is still an active volcano.

Mount Pelée, a massive volcano at the northern end of the island. Saint Pierre, a bustling city, lay at its base just four miles southwest of the summit. By April, Mount Pelée was venting sulfur from the south side of its summit and a notch had developed in its rim.

On May 5, a deadly mudslide called a lahar tore down the mountainside as the crater rim collapsed. The flow of ash engulfed a rum distillery, killing all twenty-three of its workers. With increasing unrest among the populace, the mayor, who was up for reelection, called for a scientific inquiry. A committee of the town's leaders was assembled. The only member with a science background was a high school teacher. Despite clear evidence that something ominous was in the offing, the group pronounced the city safe for the upcoming elections. The committee released its report on May 7, 1902.

Disaster struck on May 8, two days before the scheduled election. A superheated cloud of gas and vaporized rock came rolling down the mountainside, toppling stone walls like cardboard. This searing cloud of rock and hot gases blew roofs from buildings and bent iron girders as if they were taffy. The terrifying volcanic cloud was, at this point, an unknown phenomenon. It has become known as a nuée ardente (burning cloud), an appropriate term coined by the French geologist Albert Lacroix, who came to Martinique to study Mount Pelée's disastrous aftermath.

Within minutes, the center of Saint Pierre ceased to exist. The outskirts of the cloud left homes still standing, but the people who were in the houses did not survive the

On May 8, 1902, Mount Pelée erupted, destroying the town of Saint Pierre and killing all but two of its nearly thirty thousand inhabitants.

heat and gas. Tables were found with places set for a meal, with inhabitants huddled on the floor nearby. Fires burned what was left of the city. In the course of a few minutes, all thirty thousand inhabitants of Martinique's capital city had apparently perished. Remarkably, there were two survivors. The first was a shoemaker, Leon Compere-Leandre, whose house was near the edge of the nuée ardente. The searing flow burned him severely, but he recovered. After falling unconscious on his bed for about an hour, the young shoemaker awakened to find his roof burning. He escaped the newest conflagration and limped 6 km (4 mi) to the town of Fonds-Saint-Denis.

The second survivor's story is even more amazing: the victim was in the center of town at the time of the deadly glowing cloud. Louis-August Cyparis was a convicted felon. Known locally as Samson, he was in jail for attacking an acquaintance with a sword. Cyparis was doing time in solitary confinement, banished to a dungeon room with only a small air vent above the locked door. On May 8, his prison cell darkened and a gale of hot air blew into the confined room. He held his breath until the heat subsided, but the indirect encounter with the hellish nuée ardente left him severely burned on his back and shoulders. For four days he waited in terror and darkness, with no food or water. Finally, he heard the voices outside, and his cries for help were answered. After recovering from his injuries, Cyparis was pardoned. He joined the Barnum and Bailey Circus, touring the world as the "Lone Survivor of Saint Pierre."

Volcanic eruptions have caused death and devastation

The earth's tectonic plates control the location and types of volcanic eruptions on our planet.

through history, but also built new lands and made others fertile. When we think of volcanoes, we tend to remember specific places on earth: Indonesia, Italy, the western United States. How are the volcanoes distributed on our planet?

Volcanic sites follow worldwide patterns, and these patterns gave us our first inkling that the world might be segmented in a global way. Understanding the earth's plate tectonics is part and parcel to understanding its volcanoes, not only for their locations but also for their very nature. Long trains of volcanoes sweep along the floor of the oceans, tracing serpentine pathways up to volcanic chains across the continents themselves. Many of the world's volcanoes and other geothermal hotspots lie along the boundary between two plates.

The earth is a fractured world consisting of rafts of crust floating on molten rock the consistency of rubber. Plates operate like conveyor belts. In some areas, plates slide under each other, melting minerals into the magma to be recycled again by uplift (mountain building) or volcanism. At the other end of the geologic conveyor belt, new material is generated in zones of spreading. While some plates spread, others ram into each other, shoving up mountain ranges like the Himalayas and the Rockies. In other places, the plates slip and grind against each other. In these areas, earthquakes are common. The San Andreas fault of California is one such place. In still other sites, like the Andes of South America, a plate slides beneath another in a process called subduction. Subduction often leads to explosive eruptions. To many, plate tectonics was the key that finally opened a door to all of the observations we see in geology. It explained why the continents seemed to fit together, and why rare rocks from one location matched those at a distant second location. The concept of the world as a jigsaw puzzle explained widely disparate geological formations that had been mystifying.

The modern view of plate tectonics was given birth in a medical tent during World War I. While recovering from a neck wound, Alfred Wegener, a German meteorologist and geologist, came up with the theory that Africa and South America, along with all the other landmasses of the world, had once formed a supercontinent that he called *Pangaea* (*pan* meaning "all," and *gaia* meaning "earth"). He searched the world over for proof of his model, matching the profiles of mountain ranges in South Africa and Argentina, linking a plateau in Brazil with another in the Ivory Coast of Africa. He demonstrated geographic links of the fossil fern Glossopteris. Its locations line up perfectly with the coastlines of the two continents. Wegener wrote, "It is just as if we were to refit the torn pieces of a newspaper by matching their edges and then check whether the lines of print run smoothly across."

Wegener's shifting continent idea was met with widespread skepticism and outright derision. The scientific community saw Wegener's evidence as largely circumstantial. But over the years, the evidence piled up from magnetic studies of the ocean floor, orbital gravity maps, and worldwide comparisons of the fossil and geological record.

Most zones of plate-spreading lie under the surface of the ocean. Many volcanoes form along the sites of plate-spreading. The undersea mountains are hardest to study, but some plate boundaries surface here and there. The mid-Atlantic ridge stretches from north to south, nearly pole to pole. It marks a split in the world's crust that is spreading apart, making new earth every day. The ridge is essentially a mountain range towering some 1,828 to 3,480 m (about 6,000 to 11,000 ft) above the ocean floor. This remarkable ridge has a valley running down its center. This system of land-generating mountains is about 74,000 km (46,000 mi) long, the longest mountain range on earth. It is here that the North American plate moves to the west and the Eurasian plate moves east, creating new material in between. Here, too, are a multitude of volcanic sources.

Along the Atlantic ridge rumbles a chain of undersea volcanoes as alien as any we will find on other worlds. Undersea vents, grotesque chimneys of twisted rock, belch black "smoke" into the chilled water. The fragile chimneys grow quickly in delicate curtains and stony spires, some as much as 6 m (20 ft) in just a year. One such chimney, nicknamed "Godzilla," was as high as a fifteen-story building before it collapsed. The 400°C (750°F) water spewing from the submarine vents cannot boil in the high pressures of the deep sea. Bulbous "pillow lava" typical of undersea eruptions spreads across the seafloor with undulating rocky hills and ripples. Much of this volcanic territory is 3,657 m (12,000 ft) beneath the surface of the ocean, where the sandy ocean floor lies sterile in the eternal darkness. But scientists exploring this realm in deep diving submersibles discovered entire ecosystems independent of the life-giving sunlight above. Here, unlike any other biome on earth, creatures do not rely on sunlight, even indirectly. Instead, their "food source" is the chemical soup that pours from the vents. Chemobacteria in these ecosystems support a long and complex food chain of pink-gilled tubeworms that can grow to 2.4 m (8 ft), clams 30 cm (1 ft) long, one-eyed shrimp (with the single eye on its back), and blind crabs. The Pompeii worm wears a Mohawk of frilly bacteria on its back and lives in water as hot as 80°C (176°F).

One site at which the mid-Atlantic rift zone surfaces is in Iceland. Here, at the valley called Thingvellir, the land is growing at a geologically prodigious rate: some 3 mm (0.1 in) per year. Thingvellir's spectacular canyon cliffs are encrusted with moss and tundra flowers. Those picturesque walls continue into the Atlantic Ocean, where they mark the boundary of the two tectonic plates.

Plate boundaries are the drama queens of geology. In addition to earthquakes, uplift, and volcanic displays, undersea fireworks sometimes make their way to the surface. In 1963, a submarine volcano broke the surface in waters off the coast of Iceland, and the island of Surtsey was born. The appearance of Surtsey served as a window into earth's past; the island was born in the midst of molten rock and billowing steam, a scene undoubtedly common on ancient earth.

Surtsey first surfaced on November 15, 1963, off the southern coast of Iceland in an island chain called the Vestmannaeyjar (the Westman Islands). The island emanated from a fissure on the flanks of a submarine volcano. Scientists think the site erupted fairly continuously for six months before lava actually broke through the surface of the ocean. As the island built itself, fiery fountains of lava and pumice spurted hundreds of meters into the sky. The column of ash reached about 9 km (30,000 ft).

Ocean battled lava for months. The eruptions scattered cinders and lava across a growing coastline, while surf pounded the fragile beach back. A steady flow of lava finally left the island with a stable area of about 3 sq km (1 sq mi). The highest point on Surtsey reached 171 m (561 ft) above

A man stands at the edge of a rift in Iceland's Thingvellir valley where two of the earth's tectonic plates are moving apart.

sea level. Today Surtsey is a scientific reserve, a laboratory in which to study the formation of new land and its early habitation by plants and animals.

Surtsey has lots of company along the globe's plate boundaries. Volcanoes dot the mid-Atlantic ridge from the Norwegian islands Bouvet in the south to Jan Mayen in the far north. Other plate boundaries are defined by volcanic sites like the Galápagos Islands, Piton de la Fournaise on Réunion Island, and the volcanoes of New Guinea, Alaska, Siberia, and Japan. Volcano hotspots encircle the Pacific Ocean in what is called the Ring of Fire. Etna and Vesuvius tower above the place where Africa is ramming into Eurasia.

If plates are spreading apart in midocean ridges, they must be either destroyed or pushed back down elsewhere. These subduction zones occur in places such as the Ring of Fire around the Pacific. In subduction zones oceanic crust is pushed underneath continental crust and is carried steeply down into the mantle, together with sediments and water. This process generates a tremendous amount of heat. Por-

The island of Surtsey was born in 1963 when a submarine volcano broke the water's surface off the coast of Iceland. As the island was formed, fiery fountains of lava and pumice soared hundreds of meters into the sky.

tions of both the oceanic plate and the continental plate are melted and move up toward the surface. These magmas are more "evolved" (modified) than those in midocean ridges, since they are produced by melting of crust that contains sediments and other "secondhand" materials. These more evolved magmas, such as andesite, tend to produce more explosive eruptions than basaltic magmas. This leads to the exciting—and dangerous—volcanoes along the Ring of Fire such as Mount Saint Helens, the volcanoes in Central and South America and, on the other side of the ocean, those in Japan, Indonesia, and the Philippines.

Volcanoes sometimes leave evidence of the movement of earth's plates in locations far from plate boundaries. Perhaps the most famous example is tourist-friendly Hawaii, a procession of island volcanoes that trace a line roughly northwest across the Pacific Ocean. The most ancient of these volcanoes are to the north (including Niihau and Kauai), while the newest eruptions continue to form the Big Island itself, which lies nearly in the center of the Pacific

Today, the island of Surtsey is quiet. Artist-astronomer William K. Hartmann sits on the rocky ground as he paints.

tectonic plate. Beneath this slab of earth's crust, a hot source of magma pounds the underside of the plate. The hotspot doesn't move, but the plate above it does. This volcanic source periodically breaks through the moving plate, erupting into new islands. The newest and most active aboveground volcano is the Kilauea crater complex, which continues to add beachfront property to the south side of the Big Island. But beneath the surface of the ocean to the southeast, another volcano, called Loihi, is also erupting. In 2003, Loihi's summit cone collapsed, triggering submarine earthquakes similar in scope to the eruption of Mount Saint Helens. Loihi's summit is a thousand meters (3,280 ft) beneath the surface of the ocean. If eruptions continue at their current rate, Loihi will become the next Hawaiian island within ten thousand to fifty thousand years. Tourists are advised to make reservations ahead. The black sand beaches should be beautiful.

The volcanoes associated with the Hawaiian Islands tell us something else about plate movement: it is not steady

(*top*) The Hawaiian Islands, a chain of island volcanoes, trace a line roughly northwest across the Pacific Ocean. Eruptions continue to form the big island of Hawaii, which lies nearly in the center of the Pacific tectonic plate.

(*bottom*) This artist's rendition depicts Loihi (*foreground*), an underwater volcano to the southeast of Hawaii. Loihi may become the next Hawaiian island within ten thousand to fifty thousand years.

Paul Kane's 1874 painting of Mount Saint Helens shows an eruption coming out the side of the volcano. More than a hundred years later Mount Saint Helens erupted in a similar manner.

or consistent. The newest eruptions are close together and larger than older ones, implying a slowing of the plate movement. Additionally, an undersea chain of seamounts leading northward from Hawaii—more ancient volcanoes from the same stationary source—veer in a northern direction.

We see patterns of volcanic sources scattered across our maps, and there are discernible patterns in the way each volcano is expressed. Despite the fact that every volcano has its own way of doing things, volcanologists can chart similarities between them. We can learn from these patterns, but they are often difficult to read, making eruptions hard to predict. Still, lives and property might be saved if warnings gleaned from these patterns are heeded.

There is a curious painting of Mount Saint Helens in the Royal Ontario Museum. The painting, done in 1847 by Paul Kane, shows the eruption coming out the side of the volcano. The eruption of Mount Saint Helens in May 1980 unfolded in a similar manner.

Mount Saint Helens rises above pine forests in the wilderness of Washington State, about 145 km (90 mi) southeast of Seattle. Up until May 18, 1980, the majestic mountain stood nearly 2,950 m (9,677 ft) high; on this day it would lose the top 400 m (1,314 ft) of its summit—2.8 billion cu m (3.7 billion cu yd). At 8:32 a.m. Pacific Time, a strong earthquake shook the core of the mountain. The quake triggered the largest landslide in recorded history, dislodging a slurry of mud and ash that careened down the volcano's slopes at speeds up to 241 km/h (150 mph). The

(*above*) According to the U.S. Geological Survey, in December 2006 the lava dome on Mount Saint Helens was growing at an average rate of less than one half cubic meter (0.65 cubic yards) per second.
(*opposite, top*) Mount Saint Helens erupted on May 18, 1980. Spirit Lake, once surrounded by a dense forest, was in the path of the volcano's blast. Mount Rainier can be seen in the distance.
(*opposite, bottom*) More than 40,000 ha (100,000 ac) of trees were blasted flat or incinerated in the hot volcanic wind when Mount Saint Helens erupted in 1980. The trucks on the far right provide a scale.

top of the mountain was vaporized in a lateral blast that fanned out across 596 sq km (230 sq mi). Ash clouds rose to altitudes of 24,384 m (80,000 ft) in fifteen minutes, and reached the East Coast within three days. The ash would remain in the world's skies for years to come.

Although a large area around the volcano had been evacuated in anticipation of an impending eruption, experts had assumed the eruption would go up from the vent, not sideways. With the force of 24 megatons, the 349°C (660°F) blast cloud carpeted 404 sq km (100,000 acres) of wilderness with flattened trees. The number of trees destroyed was equivalent to the wood in 150,000 average-size houses.

By the end of the day, fifty-seven people had perished; twenty-one of the bodies were never found. The incredibly destructive lateral blast took most people by surprise. Had authorities been aware of the warning implicit in Paul Kane's painting, lives might have been saved. Perhaps we should have been paying attention to the personality of the mountain.

The dark areas on the moon were once oceans of lava. This image was obtained by the Galileo spacecraft on its way to Jupiter in 1992.

2 VOLCANIC PERSONALITIES

Do volcanoes have personalities? The answer is an unequivocal yes. To understand different volcanic personalities requires knowledge of volcanoes' scientific attributes. And to gain an understanding of alien volcanoes, we must first understand the workings of volcanoes on the planet we know best—Earth. We do this by comparing and contrasting terrestrial geology with that of other worlds.

VOLCANO SHAPES

The classic definition of a volcano is an opening in the earth's crust from which magma emerges. The discovery of volcanoes on other planets and moons has necessitated expanding this definition. We see volcanoes on other worlds that erupt not magma but exotic, icy mixtures that we call cryolavas. Before interplanetary exploration we could not imagine volcanoes more than 20 km (12 mi) high, plumes erupting hundreds of kilometers above a volcano, or volcanoes erupting lavas of water and ammonia.

Much of the information on the geology of other worlds comes from images, so planetary geologists compare the shape and appearance of geologic features on other planets with those on earth, to try to work out how they were formed. In the case of volcanoes, we know from terrestrial studies that such features as steep sides versus shallow sides reflect the different ways the volcanoes were formed. When a volcano—the opening in the crust—erupts, the magma forms a structure around the vent. This structure is the volcano. It is often a hill or mountain, but it doesn't have to be. Fissures can open up and spew out long lava flows, forming a sea of lava. This happened on the earth's moon. The dark patches of solidified lava on the moon were once thought to be seas of water, hence they were called "maria," Latin for seas.

Volcanoes that build up due to successive flows of relatively fluid lava piling up on top of each other are called shields. The name came from Iceland, which has many volcanoes of this type. The lava piles look somewhat like warriors' shields resting faceup on the ground. Mauna Loa in Hawaii is a good example of a shield volcano, as is Olympus Mons on Mars. Shields are often topped by large calderas created when the volcano's top opening collapses. They are truly spectacular when filled with lava lakes. Mars's Olympus Mons caldera is about 80 km (50 mi) across. We do not know how it was formed or whether it was ever filled with boiling lava. But if it were it must have been an awesome sight.

If the lavas are pastier than Hawaiian-type lavas, the resulting volcano will not descend in flat-sided slopes. Instead, the thick lava piles up, creating a dome. Domes are usually considerably smaller than shields, and their flanks are much steeper—typically 25 to 30 degrees, while the flanks of shields are usually 4 to 8 degrees. There are small domes on Mars and the moon; on Mars this type of volcano is called a tholus. A good example of a lava dome on earth is Lassen Peak in California, which last erupted in 1921.

If there is a lot of explosive activity, the resulting volcano will be neither shield nor dome. Cinder cones are typical of eruptions that have relatively mild explosions. Because the falling fragments of lava accumulate around the vent, most cones have steep and fairly uniform slopes. Sunset Crater in Arizona and Parícutin in Mexico are examples of cinder cones. Cinder cones do not appear to be at all common on other planets. Their larger cousins, stratovolcanoes, are also more common on earth than elsewhere. Stratovolcanoes are in fact the most common type of volcano on earth, but they are rare on other planets.

Stratovolcanoes are what most people think of as classic-looking volcanoes. Mount Fuji in Japan is a typical example of a stratovolcano. Others include Mount Saint Helens in the United States and Vesuvius in Italy. Stratovolcanoes are formed by the accumulation of magma fragments from explosive eruptions, interlayered with lava flows. They tend to have symmetrical shapes with graceful, upsweeping slopes. This is partly an effect of aging. When stratovolcanoes are young, their eruptions tend to come from a central vent but as they age, fractures open on their lower slopes from which lava flows come out. Explosive eruptions

MAUNA LOA VERSUS FUJI

Volcanoes come in different shapes and sizes. Hawaii's Mauna Loa (*left*) is a shield volcano, with gentle slopes that are convex-upward. Mount Fuji in Japan (*right*) is a stratovolcano, with steep slopes that are concave-upward. Though Fuji looks more impressive than Mauna Loa, it is actually smaller. Fuji rises 3,776 m (12,388 ft) over the surrounding plains; Mauna Loa's peak is 4,170 m (13,681 ft) above sea level.

The largest mountains on earth are shield volcanoes such as Mauna Loa. Measured from its base, which is on the bottom of the ocean, Mauna Loa is taller than Mount Everest. The Hawaiian volcano rises about 9,170 m (30,085 ft) from the seafloor, whereas Everest is 8,848 m (29,029 ft) high.

continue to rage at the summit and, gradually, the summit cone gets steeper relative to the rest of the volcano's body.

The type of volcanic activity clearly determines the volcano's shape and type, but what controls the type of activity? This is where the real "personality" of the volcano comes in. Volcanoes can be quiet, sending out effusions of lava without much threat to life or property. Or they can be more violent, exploding with devastating force. What causes the difference? Essentially, how much gas is dissolved in the magma and how easily it can come out. Magma has gases dissolved into it at depth, and as it moves toward the surface, the gas will strive to come out because the pressure is decreased. If the magma is viscous (pasty or squishy) but also gas poor, it can erupt quietly as stubby lava flows, often forming lava domes. Volcanologists sometimes refer to this type of lava as "toothpaste lava." Magmas that are fluid and gas-poor will produce the least explosive eruptions, though they can be quite spectacular. This is usually the case with Hawaiian eruptions. As the magma gets near the surface,

pressure from the expanding gases can spray the lava high above the vent, causing the fire fountains that make such amazing sights. We know that fire fountains have also happened on Jupiter's moon Io, no doubt in a truly spectacular way.

Magmas that have low viscosity but are gas-rich can produce explosive eruptions, but often the gas comes out gradually, and the magma will come out quietly. Some eruptions start off with only gas emissions, others produce bursts of escaping gases that can hurl magma fragments high above the ground. In general, however, eruptions from low viscosity magma are rarely violent.

The most violent volcanoes are those whose magma has both high gas content and high viscosity. This is a potentially lethal combination. Gases in high viscosity magma cannot escape easily as the magma moves to the surface. Eventually, the release of pressure allows the gases to escape explosively, tearing out the lava. Ash, gases, and hot fragments of magma erupt from the earth, resulting in some of the most deadly eruption types.

What causes these magmas to be so different? The key factors are temperature, chemical composition, the crystal content of the magma, and the amount of gas. The chemical composition of magmas on earth is, to a large extent, related to plate tectonics. In midocean ridges, basalts erupt; in subduction zones, such as the Andes, more viscous types of magma are found, such as andesite, dacite, and rhyolite. The silica in lava, combined with oxygen, forms strong bonds. Basalts have little silica, while rhyolite has the most.

(*opposite, top*) As magma approaches the surface, pressure from the expanding gases in the volcano can propel the lava high above the vent, causing fabulous fire fountains such as this one on Kilauea, Hawaii.

(*opposite, bottom*) Hawaii's Mauna Kea, an example of a giant shield volcano, is topped with small cinder cones.

(*above*) From 1969 to 1974 the Hawaiian vent Mauna Ulu, on Kilauea volcano, erupted almost continuously, often producing spectacular fire fountains. Today visitors can view the quiet crater.

The amount and size of solid crystals in the magma also contribute to viscosity. If we compare magma to cake batter, basaltic magma would be like the batter for a fluffy angel food cake, whereas a crystal-rich andesite, dacite, or rhyolite would be like batter for a fruitcake. The fluffy cake batter is easier to stir and pour. In the case of magma, the fruits are crystals, usually of pyroxene, olivine, and feldspar. These crystals can make up as much as 40 percent of the magma.

Temperature also influences the viscosity of fluids. The hotter the magma, the more easily it flows. Basalts are the hottest, melting at temperatures usually between 1000°C (1832°F) and 1250°C (2282°F). The hottest rhyolites never even reach 1000°C.

Gas content is a key factor in whether an eruption will be explosive. The composition of magma is linked to the amount of gas present, so, for example, basalts often have much less gas than the more silica-rich magmas. These differences in the magma are what ultimately cause volcanoes to have different "personalities."

ERUPTION STYLES

Eruptions on earth are usually described according to their style, and these styles have been given names. The classification, which describes how explosive an eruption is, was pioneered by G. Mercalli, who is more famous for developing the scale of seismic intensities. Most eruption types are named for the place they often occur (Hawaiian, Icelandic, Strombolian, Vulcanian); one is named for a particular volcano (Peleean); and two (Plinian and Ultraplinian) are named for the person who described a famous volcanic event (Pliny the Younger observed and described the Vesuvius eruption of AD 79). Some eruptions are described as hydromagmatic, a somewhat different eruption style that occurs when water comes into contact with hot eruption materials.

Hawaiian and Icelandic eruptions are similar: they produce long lava flows, with little explosive activity. They are often lumped together, but some volcanologists make a distinction in referring to Icelandic eruptions as the effusion of long lava flows from fissures. Hawaiian eruptions refer to lavas erupted from a central conduit to form shields. Both types of eruptions are common in Hawaii and Iceland. The trademark is long lava flows that can reach tens of kilometers and are usually thin—only a couple of meters high. Lavas tend to spread over large areas, often causing destruction to property but the mild nature of these eruptions means they usually pose little direct danger to human life. However, the indirect dangers can be great. The 1783 Laki eruption caused widespread famine in Iceland.

(*top*) Hawaiian and Icelandic eruptions produce lava flows of two types, smooth or ropy, and clinkery or rugged. They are known by their Hawaiian names, pahoehoe and a'a. The two types of lava flow on the flanks of Kilauea; the a'a is the darker lava in the background.

(*bottom*) The rugged lava known as a'a looks like a pile of jagged rocks.

Pahoehoe, or ropy lava, is smooth and relatively easy to walk on.

Hawaiian and Icelandic eruptions produce lava flows of two types, known by their Hawaiian names: pahoehoe (meaning "ropy," or easy to walk on) and the onomatopoeic a'a (sharp and rugged). Pahoehoe lava tends to form ropes on its crust as it flows and cools. A'a lava is more broken up, and the result is a flow that looks like a rubble pile of jagged rocks.

Hawaiian eruptions can form spectacular lava lakes. Relatively rare on earth, lava lakes seem to be quite common on Io. The lakes can be truly spectacular—Kilauea on Hawaii had a lava lake or pond for several years during its Pu'u O'o eruption and the Ethiopian volcano Erta Ale has had a long-lived lake.

Strombolian eruptions are named after the tiny volcanic island Stromboli in the Aeolian Sea between Italy and Sicily. Strombolian eruptions have mild to moderate explosions that look like fireworks. They have a cyclic rhythm, with explosions interspaced by rest periods that can last for seconds to half an hour or more. Volcanoes having Strombolian eruptions can stay active in this way for hundreds of years. Stromboli itself has been active for more than twenty-five hundred years. Although the characteristic activity is small explosions, lava flows can sometimes form on these volcanoes. On Stromboli itself, the flows go down a depression on the side of the volcano, known as the Sciara del Fuoco (trail of fire). The island's residents are thankful for this, as their villages elsewhere are safe from the lava.

Vulcanian eruptions are named for another island in the Aeolian Sea, Vulcano. The ancient Romans thought

VOLCANIC PERSONALITIES: SHAPES AND ERUPTION STYLES

There is a relationship between a volcano's "personality" (predominant type of eruption) and its shape.

Volcanic personality	Eruption style	Volcano shape (example)
gentle (usually has fluid basaltic magmas)	generally nonexplosive; sometimes weakly explosive	shield (Mauna Loa)
moody (magmas range from basalts to andesites)	generally, but not always, explosive	stratovolcano (Mount Etna)
violent (magmas are typically dacites and rhyolites, highly viscous)	typically highly explosive	stratovolcano (Mount Saint Helens)

that Vulcan, the god of fire, lived in the crater, where he forged armor for Mars and thunderbolts for Jupiter. And this mountain has given its name not only to a type of eruption but also to all the volcanoes in the solar system. Vulcanian eruptions are more violent than Strombolian ones, as the magmas are usually more viscous. Typically, this type disgorges vast quantities of ash, which can rise more than 10 km (6 mi) into the atmosphere and cause ashfall over large areas. The ominous black eruption clouds can disrupt flights, affect weather, and upset day-to-day living. The Sakurajima volcano in Japan has been erupting in this fashion since 1955, causing the residents of nearby Kagoshima a great deal of concern and discomfort.

The next category, Peleean eruptions, was named after Mount Pelée in Martinique. These eruptions can be nasty indeed. When Mount Pelée erupted in 1902, some thirty thousand people died (see chap. 1). The two hallmarks of Peleean eruptions are pasty lava domes and the deadly pyroclastic (fragmented rock) flows, also called nuée ardentes (glowing clouds). Nuées have been responsible for many of the deaths caused by volcanoes. Pelean eruptions are very dangerous. A modern example is the 1951 eruption of Mount Lamington in Papua New Guinea, which killed three thousand people. The people had not known that Mount Lamington was an active volcano.

The most violent types of eruption are Plinian and its bigger cousin, Ultraplinian. The most famous example of an eruption of this type is Vesuvius in AD 79, which destroyed the towns of Pompeii and Herculaneum. Plinian eruptions

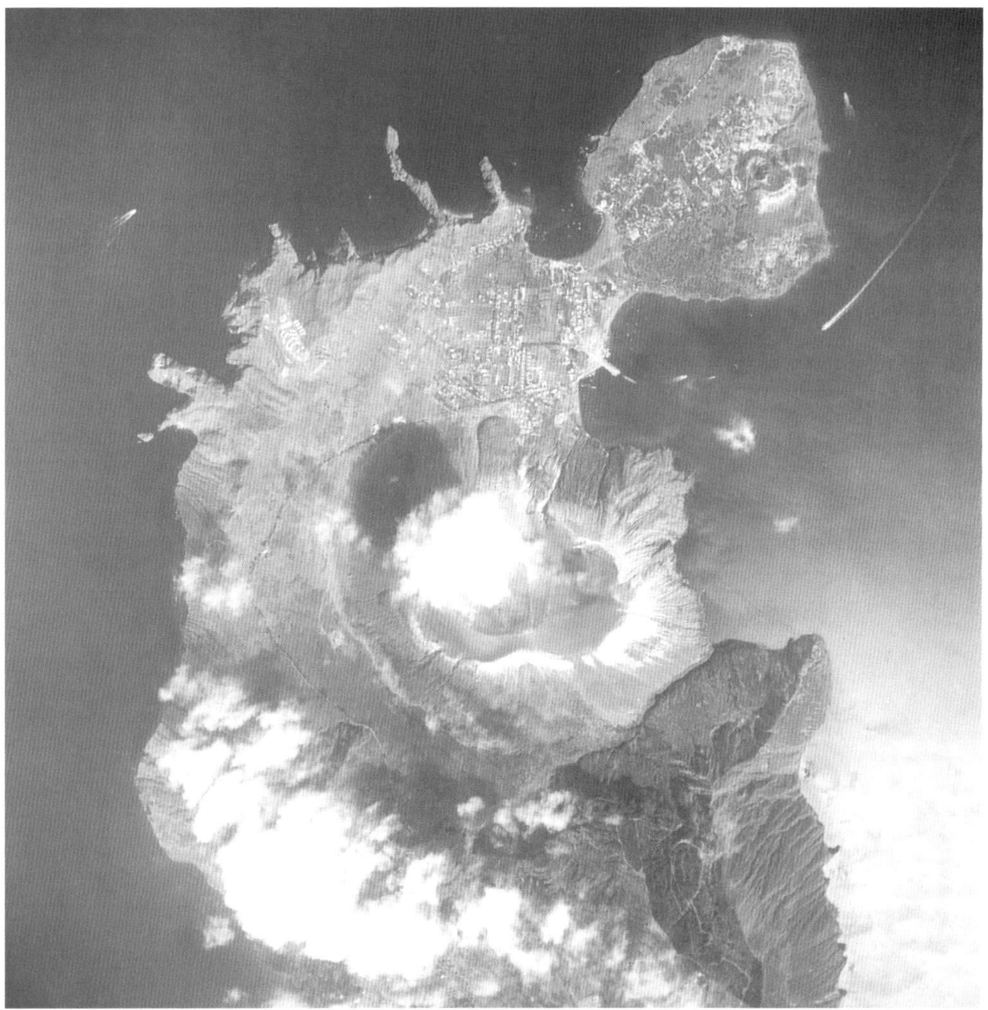

have extremely violent explosions that eject great volumes of ash into the atmosphere. The eruption columns can reach up to the stratosphere and disperse ash over extremely large areas. The effects can be felt all around the globe, in the form of red sunsets and a drop in average temperatures. Plinian eruptions often release pyroclastic flows and sometimes another deadly type of flow—a lahar, or mudflow. Lahars occur when large quantities of ash mix with water from rainfall or melted ice. The consistency of wet cement, the mixture can speed down the volcano, burying everything in its path.

The most destructive eruptions in history were Plinian and Ultraplinian. The Taupo volcano erupted in New Zealand in AD 186. Fortunately, this cataclysmic event occurred before settlers arrived, but Taupo could put on a repeat performance. The 1883 eruption of Krakatau is one of the most famous Plinian eruptions (see chap. 1), but another Indonesian volcano produced even more dramatic results. In 1815 Tambora spewed forth some 40 cu km (10 cu mi) of ash and magma fragments. Ash and aerosols were injected into the earth's stratosphere, dramatically affecting the climate. In Europe, 1816 was known as "the year without a summer." The dramatic cold snap inspired Lord Byron's poem "Darkness" and even, in a roundabout way, a renowned novel. Byron and two friends endured such gloomy weather during a vacation that summer that he suggested they amuse themselves by writing ghost stories. One of the friends was Mary Shelley, who penned the most famous horror story of all time—*Frankenstein*.

(*opposite, top*) The Stromboli volcano in Italy forms a tiny island in the Aeolian Sea. The flows go down a depression on the side of the volcano; it is called the Sciara del Fuoco (trail of fire).

(*opposite, bottom*) Vulcano, the southernmost of the Aeolian Islands, seen from above. It is the volcano from which all others get their name.

(*above*) A lava fountain erupts from Kilauea, Hawaii. Hawaiian-type volcanoes generally have eruptions that are not explosive but are spectacular to watch.

Among the most dramatic volcanic forms on Venus are the "pancake domes." These steep-sided domes are formed by thick, sticky lava flows. Here, an artist imagines two such volcanic mountains rising from the plains of Venus as seen from within a fracture on the surface that cuts across the right dome in the distance.

3 EARTHLIKE WORLDS

Which planets other than earth have volcanoes? The question has long been in the minds of scientists and artists. In September 1947 the artist Chesley Bonestell depicted the surface of Jupiter for *Air Trails and Science Frontiers* magazine. The scene portrayed an immense volcano birthing a Yosemite-esque lavafall. A later version was captioned "Hydrogen flames and lava pour out off top of cliff. Lake below is liquid ammonia; cliffs are lava and ice." The science writer Willy Ley, in his Bonestell-illustrated book *Conquest of Space*, commented on the mottled jumble of clouds, ovals, and belts in Jupiter's atmosphere: "These disturbances may be called 'volcanic,' although they are more likely to be chemical explosions of hydrogen, caused possibly by sodium or other chemical reactions which are not in our low-pressure chemistry."

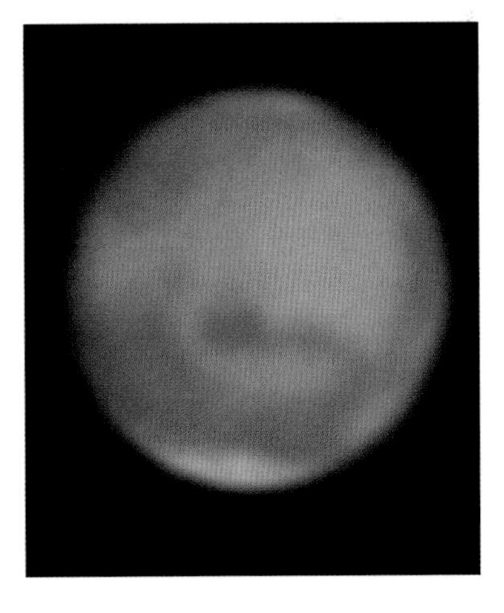

Very few details are visible in early telescopic views of the Red Planet.

Ley had no idea just how alien the chemistry and environment of Jupiter really was. Today, we know Jupiter is a world with no solid surface. Its core is denser than the center of the earth, but its density gradually dissipates, changing from solid core to liquid atmosphere to gas. Jupiter has no volcanoes, as it has no solid surface to put them on. Bonestell, who is considered the father of modern space art, accurately reflected scientific theory of the time. Fuzzy telescopic views and ambiguous remote data were the only means available for determining where alien volcanoes might exist and what they might be like. Jupiter was only one of the candidates. Did billowing cones enrich jungles under the fog of Venus? Were fire fountains illuminating

Voyager 2 captured this image of Jupiter's great red spot in 1979.

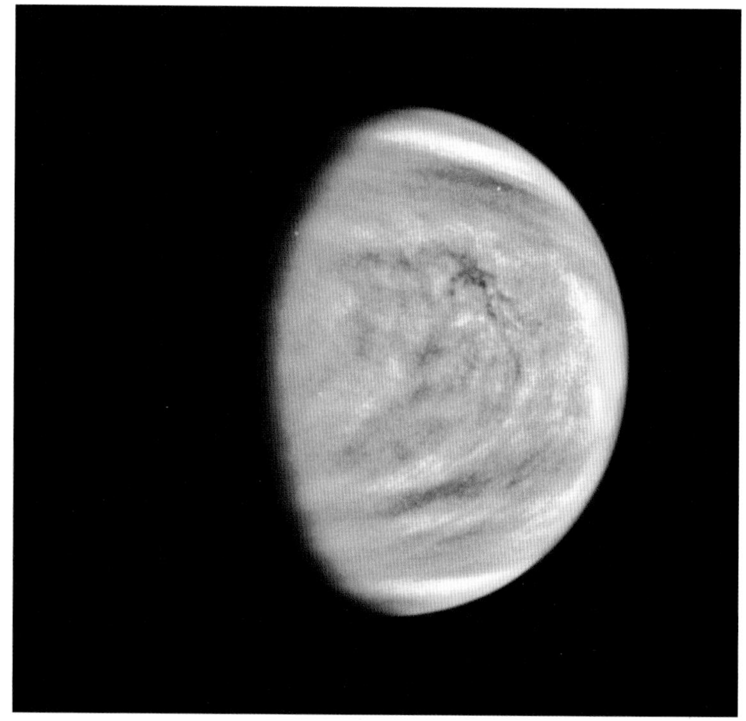

This image of Venus was taken by the Galileo spacecraft in 1991.

Chesley Bonestell, who is considered the father of modern space art, envisioned Jupiter as a world with volcanoes. In fact, the planet has no solid surface and thus no volcanoes.

the long nights on the far side of the moon or spraying their golden embers across desert floors on Mars? Perhaps volcanoes were responsible for the rings around Saturn or for the scattered debris of the asteroid belt. Space exploration has enabled us to view other worlds up close and to find strange and beautiful volcanoes on them, often in unexpected places.

MARS

Volcanic mountains—the type of volcano many people think of as "real volcanoes"—were not conclusively discovered on other worlds until 1971, when NASA's Mariner 9 became the first spacecraft to orbit Mars. The Red Planet held much promise as a place with earthlike deserts, canals, and maybe even life. Mariner 9 began circling the Martian globe at the climax of a planetwide dust storm. Astronomers knew of these storms from centuries of telescopic observation. Seasonal variation in temperature triggers dust devils and local dust storms, which often combine to form planetwide maelstroms. Two months before Mariner 9 dropped into orbit, surface winds up to 180 km/hr (112 mph) kicked a billion tons of talcumlike dust into the Martian sky.

Mariner 9's ringside seat for the storm of '71 may have been exciting for planetary meteorologists, but it was disappointing for a world waiting to see the real Mars below. As the dust began to settle, a dark object appeared in an area known as Nix Olympica (the Snows of Olympus, named after bright transient areas observed by telescope). At first,

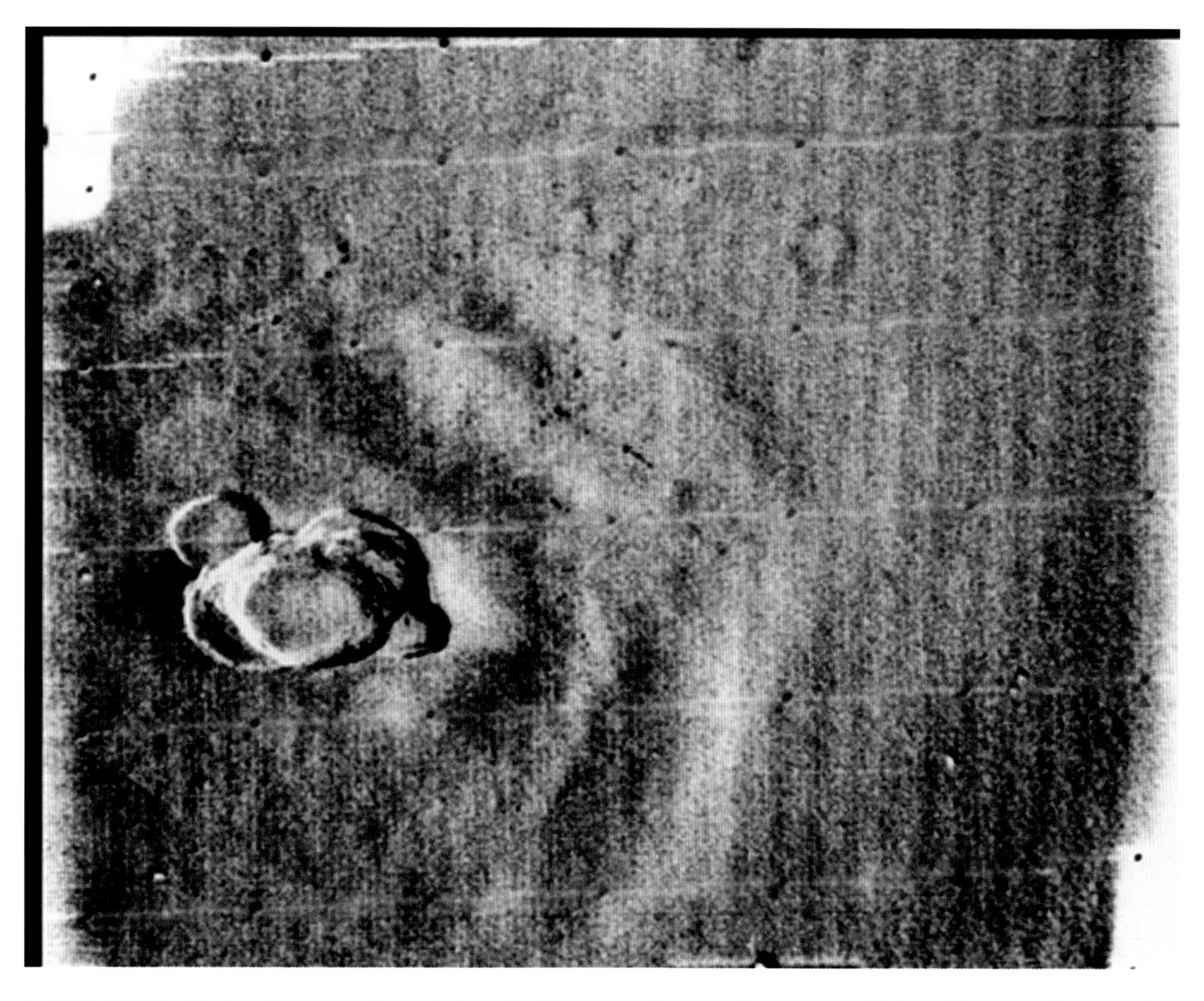

In 1971 NASA's Mariner 9 spacecraft settled into orbit around Mars. In this image, volcanoes can be seen rising out of a global dust storm.

the flat-topped summit was mystifying. More peaks reared out of the dust as the storm abated further, revealing features reminiscent of volcanic calderas in Hawaii.

As mapping of Mars continued over the next eleven months, cartographers rechristened Nix Olympica as Olympus Mons (Mount Olympus), for it is the largest mountain we know of in the solar system. It's also the largest known volcano, with a summit caldera standing 24 km (15 mi) high, nearly three times as high as Mount Everest. The massive shield volcano's base, about 600 km (373 mi) across, would nearly cover the state of New Mexico.

The northwest flank of the volcanic province Tharsis is home to the largest of Martian volcanoes. Arsia Mons, Pavonis Mons, and Ascraeus Mons form a towering triumvirate, with summits 18, 14, and 18 km (about 11, 9, and 11 mi) high, respectively. North of these monstrous mountains lies Alba Patera, an ancient, 1,500-km-wide (932-mi) volcanic structure scored by a web of branching valleys.

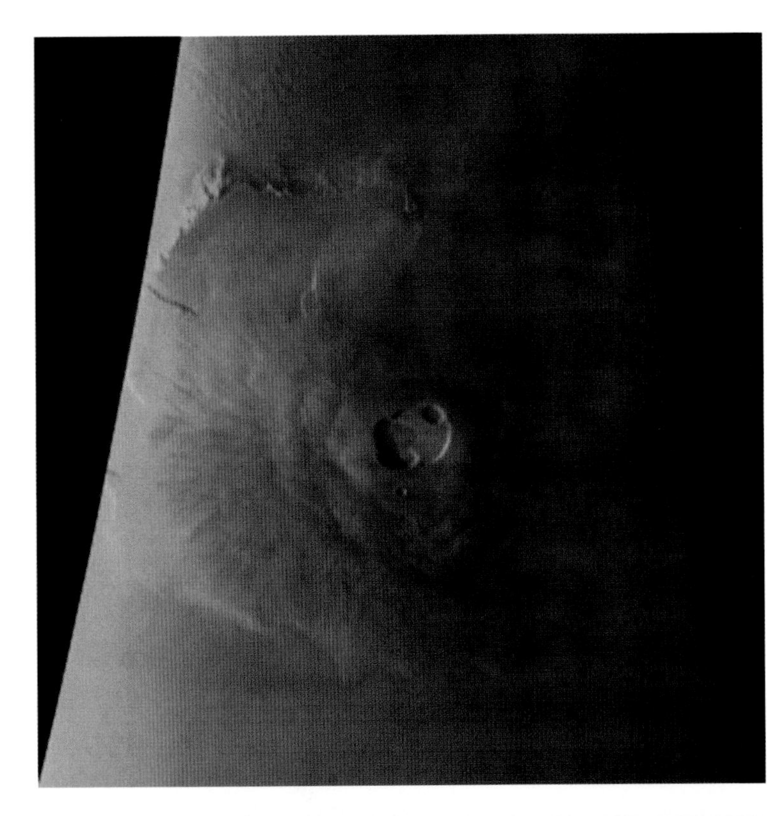

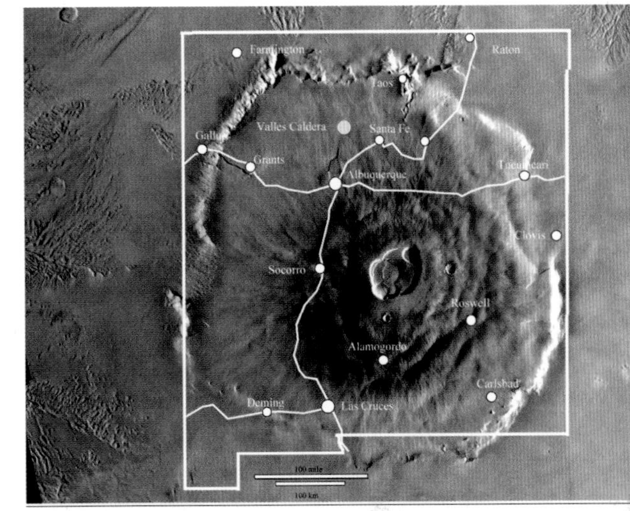

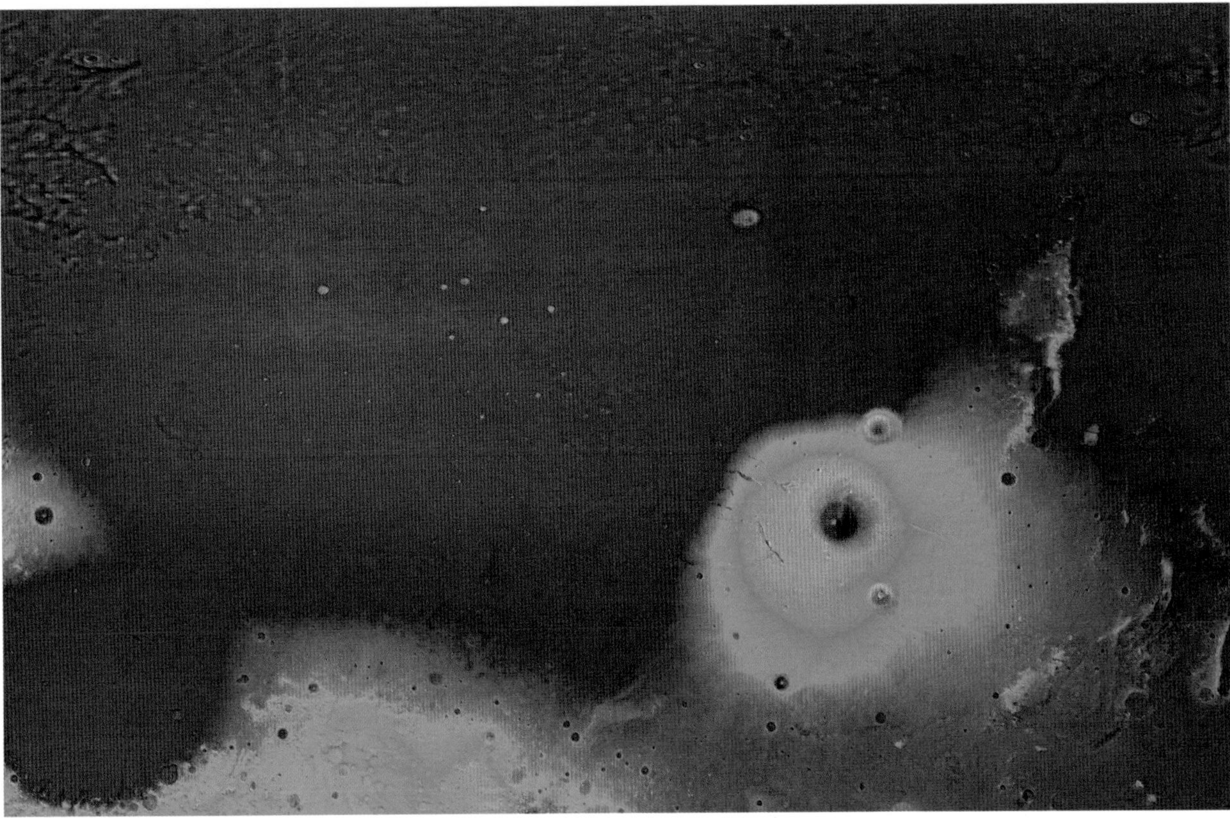

(*top, left*) Olympus Mons, the largest known mountain in the solar system, is nearly three times as high as Mount Everest.

(*top, right*) Olympus Mons, which is 600 km (373 mi) wide at its base, would cover almost all of New Mexico.

(*bottom*) In Greek mythology, Elysium was a place of peace and harmony. The Martian Elysium may be a peaceful place now, but in the past it was wracked by volcanic activity.

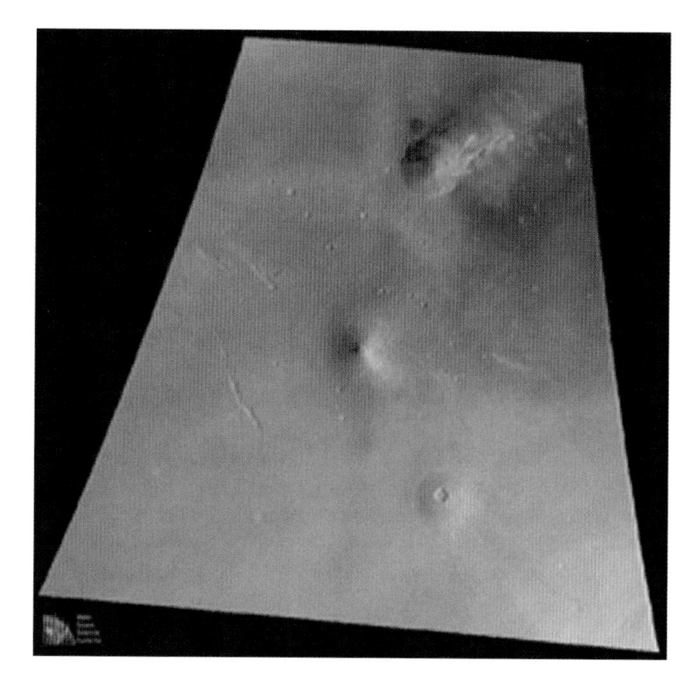

Elysium Mons, at the center of the Martian Elysium, rises 9 km (nearly 30,000 ft) above the ancient lava plains. It is the fourth tallest mountain on the planet.

Alba is capped by a caldera 100 km (62 mi) in diameter. The slopes of the mountain fan out to cover eight times the base area of Olympus Mons. Alba is covered by radial ridges that maintain a consistent width over long distances. These ridges are typically crested by a channel or series of pits, implying subsurface lava tubes. Smaller cones and vents break the surface on the flanks of the mountain. Superimposed on the lava flows is a network of branching valleys. Many of these dendritic (branching) forms wander down the mountainside with the same characteristics as water or mud flows. Alba exhibits many telltale signs not only of volcanism but also of water erosion.

Nearly on the opposite side of the planet from Tharsis is another major volcanic region known as Elysium. In Greek mythology, Elysium was a place of peace, harmony, and rest. The Greeks believed that Elysium lay on the western edge of the earth, perched on the western beach of the great Oceanus, a waterway encircling the entire world. The Elysian Fields afforded a final resting place for the souls of heroes.

The Martian Elysium may be a peaceful place now, but it was once wracked by violent volcanic explosions. The many volcanoes in Elysium are substantial in scale. At the center, Elysium Mons rises 9 km (nearly 30,000 ft) above the ancient lava plains, making it the fourth tallest on the planet. The province of Elysium Planitia builds some 5 km (16,404 ft) above Martian "sea level," giving Elysium Mons's summit an altitude of 14 km (45,932 ft). The mountain is unique in form. The summit caldera sits atop an asymmetri-

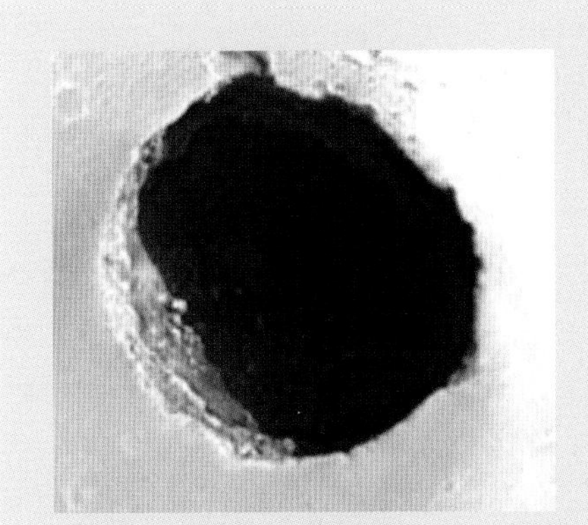

MARTIAN CAVES

Mars's volcanic provinces may be peppered with caves. High resolution orbital images have revealed features on several Martian volcanoes that appear to be openings into lava tubes. Several of these "skylights" are as wide as a football field and lead to some kind of internal cavity. Dubbed the Seven Sisters, the caves were first noted as black spots on the flanks of Arsia Mons. The sites have since been resolved into portals over underground chambers.

Another location has an archway left behind by collapsing ceilings in a lava tube. The lava tube, about 24 m (79 ft) across, snakes between hills in the Tartarus Colles region of Mars. Since caves are natural collectors of water vapor, these subterranean doorways may be a future target in the search for past or present life on Mars.

cal circular mountain. To the north, a distinctive 2-km-high (6,562-ft) bench extends some 200 km (124 mi). This extension is draped by flows and sinuous ridges reminiscent of some areas in Tharsis, but the main mountain has no such flows. It is covered by hummocky terrain. Surrounding the entire structure are concentric fractures. These may be the scars left after the Elysium volcanoes rose, their weight bearing down on the Martian crust until it cracked.

Judging by crater counts, Elysium Mons is one of the older large volcanoes on Mars, certainly older than the Tharsis shields. Its flanks have been silent for at least 600 million years. But nearby flows tell a different story. The Elysium region is blanketed by some of the youngest lava flows on Mars. In many places, outflow channels score the flanks of volcanoes. Titanic floods have sculpted these valleys, often pouring forth from fractured or collapsed terrain. It is clear that Martian volcanoes have played a critical role in the origin of its water-carved features. The flanks of Elysium plateau are inundated with the remains of channels

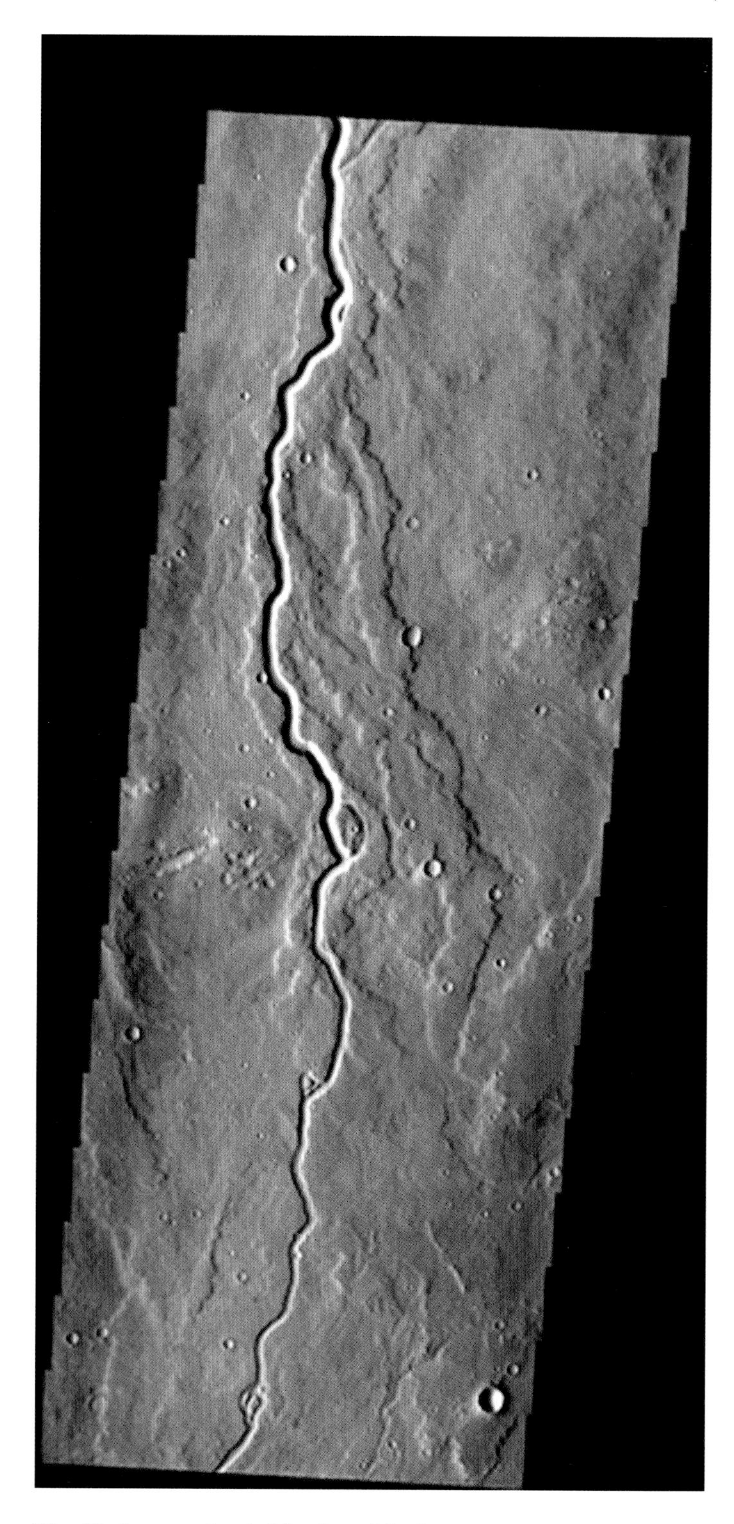

The Elysium region is blanketed by lava flows. This image of a lava channel on Elysium Mons was taken by the Mars Odyssey mission.

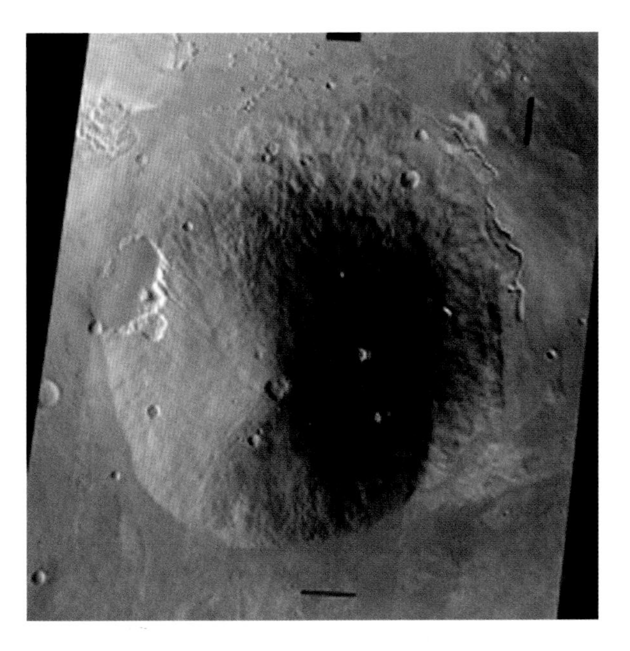

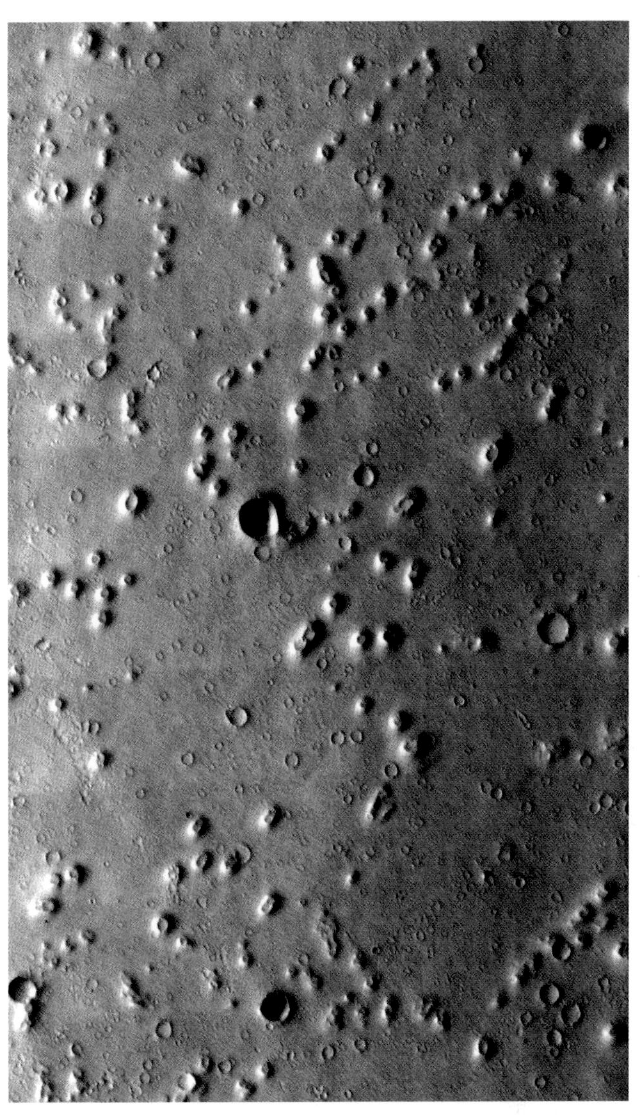

and twisting mudflows. Many researchers think it likely that geothermal sources have melted permafrost or frozen ground water, triggering outbursts of water and mud.

Shields are not the only type of Martian volcanoes. Elysium's Hecates Tholus is a prime example of a steep-sided volcano thought to have been formed by explosive activity. Its summit is crowned by a complex caldera 11 km (about 7 mi) across. The caldera records multiple collapse events, and its rim has been breached in at least two places, with the breaks extending down the mountain in sinuous channels. Several graben and crater chains radiate down the flanks. Most of the craters appear to have volcanic—rather than impact—origins. A dramatic embayment slices into the western slopes of Hecates Tholus.

Volcanic domes and cinder cones are another common feature on Mars. Across the ochre plains of Elysium, hundreds of crater-capped mounds punch through the rolling plains. Some exhibit the classic profiles of domes formed by the interaction of lava and ice. Cinder cones and domes

(*left*) Elysium's Hecates Tholus, a steep-sided volcano thought to have formed by explosive activity, is crowned by a complex caldera 11 km (7 mi) across.
(*right*) The pock-marked surface of Isidis Planitia on Mars may be the result of volcanic activity.

(*opposite*) **In this artist's rendering, a volcanic eruption triggers flash flooding across the Martian plains as an ice crystal halo forms in the sky.**

(*above*) **This image of Tyrrhena Patera, a volcano in the southern highlands of Mars at the edge of the Hellas impact basin, was taken during NASA's Viking mission.**

rise from many sites across the Red Planet. In Isidis Planitia, hundreds of small cones (about 500 m [1,640 ft] in diameter) march across the plains in serpentine lines. They may be spatter cones defining the same fissures that gave birth to the lava plains of Isidis.

Other domes, most with slotted vents at the crest, scatter across the uneven landscape east of the great Hellas impact basin. Hellas is the largest identifiable impact ring on Mars, spanning 2,092 km (1,300 mi) across. Its outer ring rises 2 km (1 mi) and stretches nearly 4,023 km (2,500 mi) from the center of the basin, which is nearly 10 km (6 mi) deep. Remnants of volcanoes found within this ring are among the planet's most ancient.

One of the finest examples of volcanoes in the region is Tyrrhena Patera. The site is spectacular, not for its height—which is a modest 2 km (1 mi)—but for the remarkable flows and furrows radiating across its slopes. This highland volcano is a 300-km (186-mi) disk of lobate sheets of material. The mountain was probably built more of ash than of lava flows. Most of its eruptive events may have been hydromagmatic (or phreatic), a style of eruption that occurs when water comes into contact with hot eruption materials.

Direct evidence for the nature of Martian volcanism has come from the Mars Exploration Rover Mission, which launched the rovers Spirit and Opportunity in the summer of 2003. Spirit's landing site inside the Connecticut-size Gusev Crater yielded extensive evidence of volcanism in samples from the crater floor as well as sites up in the Columbia Hills. One unusual rock found early in the mis-

Dramatic flows emanate from the Martian volcano Tyrrhena Patera.

sion was a dark volcanic rock about 70 cm (2 ft) tall. NASA personnel nicknamed it Humphrey. Within the rock, bright cracks contained minerals that likely crystallized out of flowing water. The water may have been mixed into the original magma that formed Humphrey, or it may have flowed through the rock later, after it hardened. Spirit sampled rocks ejected from an impact crater dubbed Bonneville, confirming that volcanic rock extends at least 10 m (33 ft) below the surface of Gusev's floor.

Spirit labored through 3 km (1.8 mi) of Martian sand to reach the Columbia Hills. There, the MER (Mars Exploration Rover) focused its instruments on a rock known as Wishstone. Wishstone exhibits forms that may have resulted from an explosion, presumably a meteor impact or volcanic event. Another rock, nicknamed Peace, contained olivine, pyroxene, and magnetite, all of which are common to volcanic rock. The layered rock that makes up the Columbia Hill assembly in Gusev resembles rock formed by volcanic ashfall or ash flow, both in form and in chemical makeup. In short, the rovers have uncovered significant traces of volcanism.

How old are the volcanoes of Mars? Geologists separate Martian history into three eras: Noachian (the oldest, from the planet's formation up until roughly 3,500 million years ago), Hesperian (from 3,500 to 1,800 million years ago), and Amazonian, which reaches into the present. The Hellas basin's Tyrrhena is thought to have first erupted roughly 3,000 million years ago and probably continued activity into the early Amazonian period, up to 1,600 million

(*top*) The Spirit rover captured this view of a dark boulder about 40 cm (16 in) tall, one of the many dark, volcanic rock fragments scattered on the slope of "low ridge." The rock surface looks similar to those found on the outside of lava flows on earth.

(*bottom*) This group of basaltic rocks was informally named FuYi. In ancient Chinese myth, FuYi was the first great emperor. He explained the theory of yin and yang to his people and was the first person to use fire to cook food.

(*opposite*) **A distant eruption triggers mud flows as underground ice melts in this artist's rendering of ancient Mars.**

years in the past. Volcanoes from Mars's earliest formative periods have been all but obliterated by meteor impacts and relentless winds. Studies of the rare Noachian areas still preserved show subtle evidence of volcanism, and some scientists suggest that Martian volcanism was ten times as fierce then as during the Hesperian era. But volcanic styles have changed over time. The earliest Martian environments were probably wetter than they are today. Water tends to cause explosive eruptions, forming more ash than lava, and ash-built structures tend to weather away faster than rugged lava mountains. The large shields or other volcanoes that did build up tended to sink into the crust under their own weight. Therefore, the shields of Tharsis and Elysium are fairly recent, with their activity continuing into the Amazonian. In fact, some flows on the Tharsis shields could have been laid down within the last 10 million years—quite recent in geologic terms.

Evidence for recent volcanism comes from different sources. One of these involves analysis of rare Martian meteorites. Some of the most recent Mars rocks crystallized out of lavas about 170 million years ago. They were blown off of the Martian surface by a huge meteor or asteroid roughly 3 million years ago. Based on their composition and structure, their history is well understood. The meteorites were born in magma more than 7 km (4 mi) below the Martian surface, where they underwent some cooling in a subterranean chamber. They were then erupted onto the surface of the planet in a lava flow that was at least 10 m (33 ft) thick. Researchers know this because the rocks contain crystals that form only during slow cooling (the kind that takes place in a thick pile of lava). The types of rocks represented by these meteorites are common—and well understood—here on earth. Only a careful study of the gases and isotopes within the meteorites told scientists where these rocks came from. What is more significant is the *when*: these chunks of basaltic lava came to the surface in what was essentially a modern eruption. According to this evidence, volcanism on Mars was occurring in the recent past.

Another line of evidence concerns fissures in the Cer-

(*right*) The Allan Hills meteorite, which was found in Antarctica in 1984, displays evidence of Martian volcanism and flowing water. Some scientists think that the discovery of tiny magnetic crystals inside the 4.5-billion-year-old rock is an indication that life once existed on Mars.
(*below*) Platy flows in Marte Vallis clearly indicate volcanic activity.

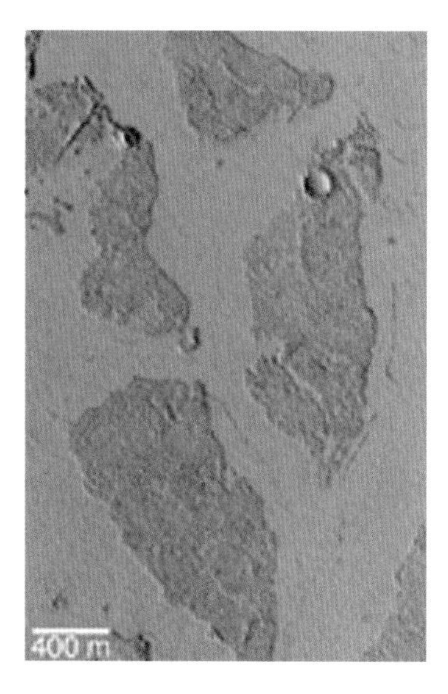

berus region: they seem to be leaking dark material onto the surface. MGS (Mars Global Surveyor) images do not show enough detail to discriminate between lava flows or mudflows, but Cerberus canyons show some kind of recent geologic activity. Portions of the Cerberus fissures began as a series of pits, another indication of volcanic-related formation. Are these fissures still active today? Flows in other places are platy, clearly volcanic in nature.

To some researchers, the most tantalizing circumstantial evidence for active Mars volcanics comes not from orbiters, landers, or meteorites, but rather from remote sensing here on earth. In September 2003, a NASA team reported on research carried out at the Gemini South telescope on Cerro Pachon, Chile. Observations there were combined with data from the Keck II telescope on the summit of Hawaii's Mauna Kea. The telescopic observations, which were taken in the infrared spectrum, indicated that a very small amount of methane was present in the Martian atmosphere. The trace amounts—about 10.5 parts per billion—were detected by sensitive infrared spectrometers attached to the Gemini and Keck instruments. Because methane is an unstable gas, the estimated lifetime of methane on Mars is estimated to be only a few centuries. For methane to be detected in the atmosphere, it must be replenished continually, and a common agent of replenishment is volcanism.

Hot on the heels of the NASA team's discovery came a report from a Catholic University of America team. Using the Canada-France-Hawaii telescope, the group found methane absorption lines in the Martian spectrum. Their results

NASA's Mars Science Laboratory (*left*) is designed to collect Martian soil and rock samples and analyze them for organic compounds and environmental conditions that could have supported microbial life now or in the past. Europe's ExoMars lander (*right*) will search for signs of life on the Red Planet.

were in turn buttressed by the European Space Agency's Mars Express orbiter. Mars Express, which carries a Fourier transform spectrometer, took nearly seventeen hundred samples from orbit. The three independent studies bolster the assertion that something is pumping methane into the Martian atmosphere. But what is it? Volcanism is a possibility, but if volcanoes are the cause, where are the vents? Why haven't they been found in the infrared spectra of the orbiters or ground-based telescopes? Perhaps it is simply because resolution has not been good enough to site small geothermal hot spots.

Exobiologists suggest another possibility for methane in the Martian atmosphere: biological activity. Methane-producing microbes, which are well known on earth, would not be visually detectable, even through the eyes of landers such as the MERs or the Viking and Pathfinder landers. More sensitive biological sensors are slated to fly in future missions aboard NASA's super rover, the Mars Science Laboratory, and Europe's ExoMars lander.

While the search continues for active volcanism on Mars, some researchers think there is a better chance of finding active eruptions on earth's other next-door neighbor. Venus is slowly giving up its secrets.

VENUS

Among the terrestrial (earthlike) worlds, Venus is Planet Volcano. Venusian real estate has more volcanic shields, domes, cones, and flows per square mile than any other planet in our solar system. Were it not for the gloomy, acid-laden

Venera 13 landed on Venus in March 1982. It survived on the surface for 127 minutes during which time it transmitted data, including this image of the planet's surface.

yellow sky, some of its volcanic landscapes would seem familiar to any Hawaiian. Other landforms are alien beyond the imaginings of the best science fiction writers. Their bizarre nature may stem from the Venusian environment. Surface temperatures simmer at more than 700°C (900°F). Air pressure at the surface is ninety times that of earth at sea level, something more akin to the depths of earth's oceans. Sulfuric acid drizzles in the lightning-laced clouds above. It's not a great place for a picnic.

Many Venusian volcanoes are in pristine condition, seemingly formed only yesterday. And they're everywhere; at least 90 percent of earth's twin is wrapped in volcanic features. The surface of our cloud-covered celestial neighbor was revealed, bit by bit, through the radar eyes of Russian Venera and U.S. Pioneer and Magellan orbiters. As the fog lifted, familiar formations appeared. For example, earth's ocean basins find their counterpart in the lowlands of Venus. The Venusian oceans—devoid of water—had flowing liquid in the past; tides of molten basalt broke upon Venusian "coastlines." Flash floods of extended lava flows fanned out into smooth flatlands. Remnants of these hellish days reveal themselves in long, sinuous channels that snake their way to the rock oceans. Some of them are thousands of kilometers long. The Baltis Vallis may be the longest lava channel in the solar system, 1.5 km (1 mi) wide at places and more than 6,437 km (4,000 mi) long. Both ends of the dried-up lava channel have been obscured by newer flows; it may have been considerably longer.

The long lava flows have left features comparable to

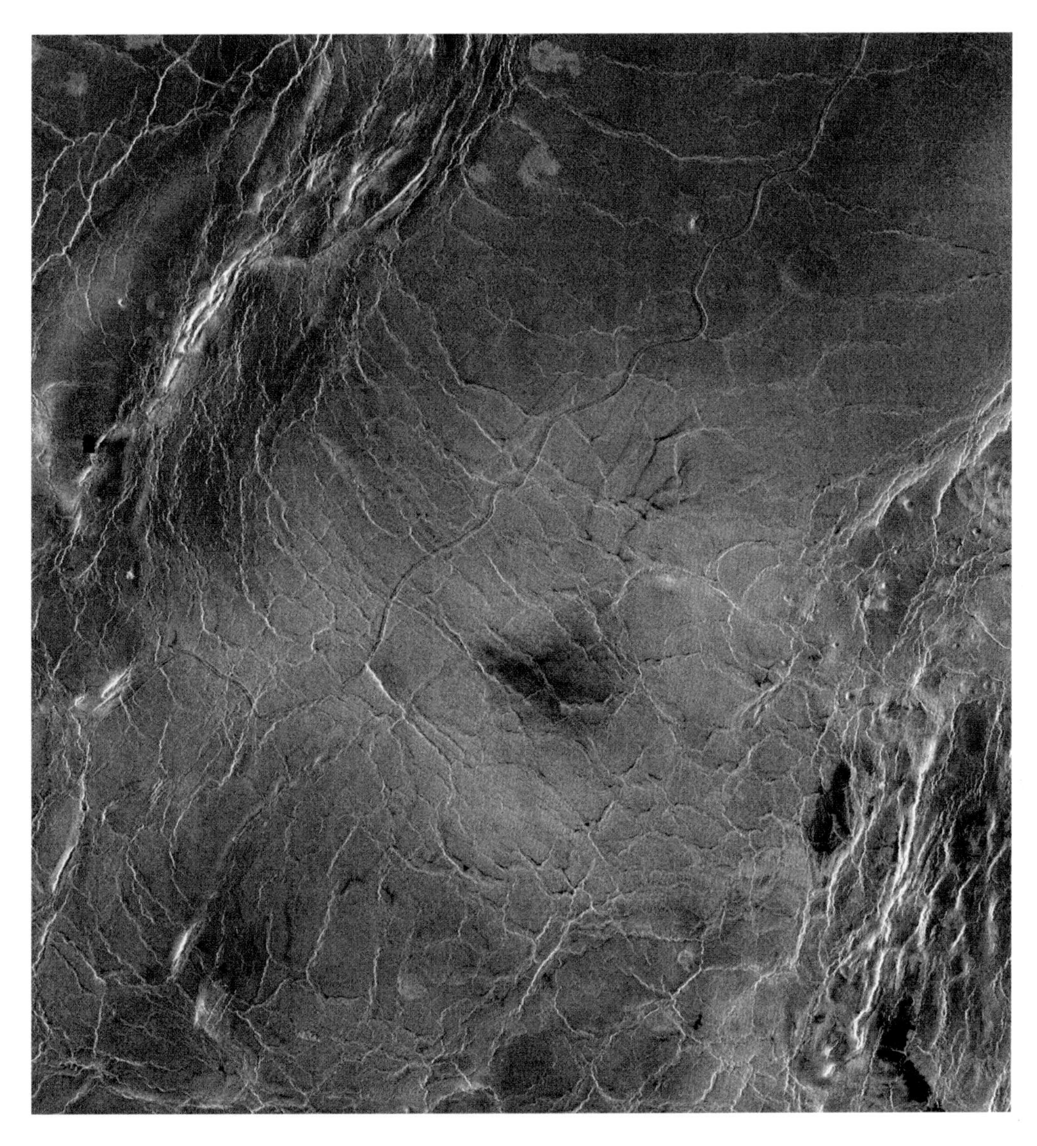

terrestrial rivers: bars, oxbows, and meanders. Many fan out in the end as classic deltas. In some places we find evidence of shifting in the lava's flow and migration of floodplains. These features are characteristic of extended, long-term processes rather than flash floods or single-event eruption flows. Apparently, something in Venusian chemistry allows lava to flow over long periods and long distances. Some sulfur-rich types of rock can stay molten and thin for very long periods of time. Ultrabasic lavas, such as those that may be present on Io, are another possibility, as they have very low viscosity and can flow long distances. Another possibility, carbonatite, was suggested by David Grinspoon and others. Carbonatite volcanism, which is extremely rare on earth, involves carbon-rich rock with a very low melting temperature, 490°C (914°F). This is close to the Venusian daily temperature. Some researchers even envision a carbonatite aquifer stored below the surface, serving a similar role to earth's underground water table. For now, the long-lived lava flows of Venus remain a mystery.

More than a kilometer (0.6 mi) wide at places and more than 6,347 km (4,000 mi) long, the Baltis Vallis on Venus may be the longest lava channel in our solar system.

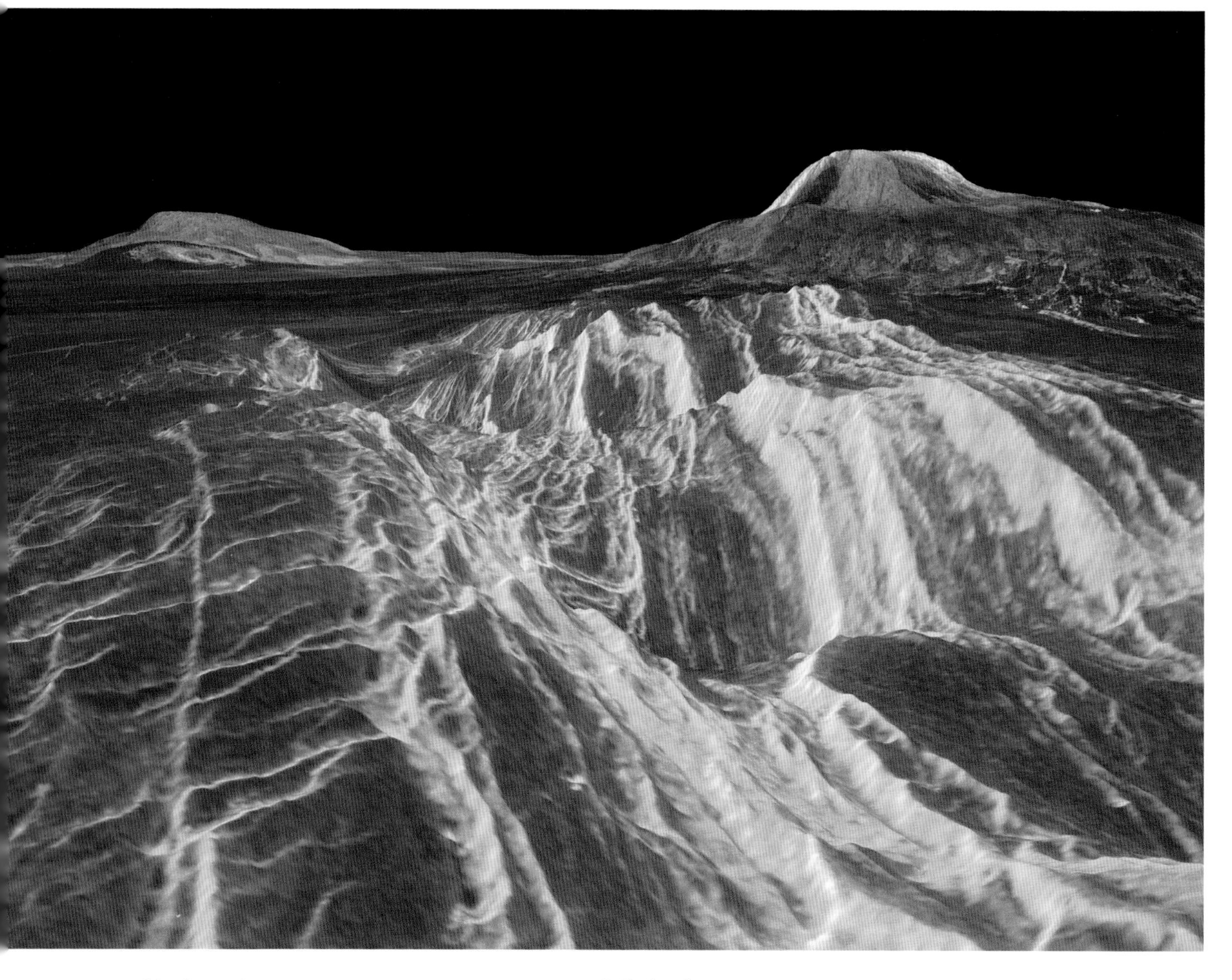

This three-dimensional computer-generated view of the surface of Venus shows two volcanoes on the horizon. On the right, Gula Mons reaches 3 km (2 mi) high with a diameter of 400 km (249 mi). Sif Mons, on the left, has a diameter of 300 km (186 mi) and a height of 2 km (1 mi).

While the composition of the lavas on Venus remains unknown, evidence of eruption is everywhere. Among the volcanoes most familiar to us earth-dwellers are the shields. Scattered across the Venusian lowlands, most are 20 km (12 mi) or less across. Like the Hawaiian shields, these gently sloped summits are the byproduct of many intermittent flows of thin lava. Eons of gradual buildup result in these low-profile mounts.

In many areas, the volcanoes congregate in what researchers call "shield fields." These clusters average roughly 161 km (100 mi) across. The vast ocean plains of Venus may play host to hundreds of thousands of small shield volcanoes, a number similar to those found on the earth's ocean floor.

Some Venus shields have grown to gigantic size, towering 7,924 m (26,000 ft) into the sulfuric acid haze. They dot the plains and rest atop the highlands. Emblematic of these giants is Gula Mons. It rises in the northwest region of Eistla Regio, a 6,000-km-long (3,728-mi) east-west trend-

ing upland area that abuts the great continent of Aphrodite Terra. The mountain towers about 5 km (3 mi) above Eistla. Rugged scarps and troughs score its northern face, radiating out from the central summit caldera. Magellan radar images indicate that the northwestern face is very rough. The area may represent the most recent of lava flows and be similar to a'a flows on earth. A rift valley extends from Gula Mons's base, indicating that the Eistla region has been subjected not only to volcanism but also to faulting in a geologically active past.

While the shield volcanoes of Venus may conjure up thoughts of tropical black sand beaches, other sites have no earthly analogy at all. Perhaps most remarkable are the coronas. Coronas are circular or oblong systems of fractures and wrinkles, blemishes on the Venusian crust spanning hundreds of kilometers. A trough usually circumscribes the ringed fracture system of the coronas. Nothing like them has been found on any other planet or moon. The mother of all coronas is Artemis Corona. While most coronas are

Aine Corona, the large circular area near the center of this Magellan radar image, is approximately 200 km (124 mi) in diameter. Two "pancake" domes are also visible. One, just north of Aine Corona, is about 35 km (22 mi) in diameter. Another is located inside the western parts of the annulus, or ring, of the corona fractures. Complex fracture patterns like the one in the upper right are often seen in association with coronas and various volcanic features.

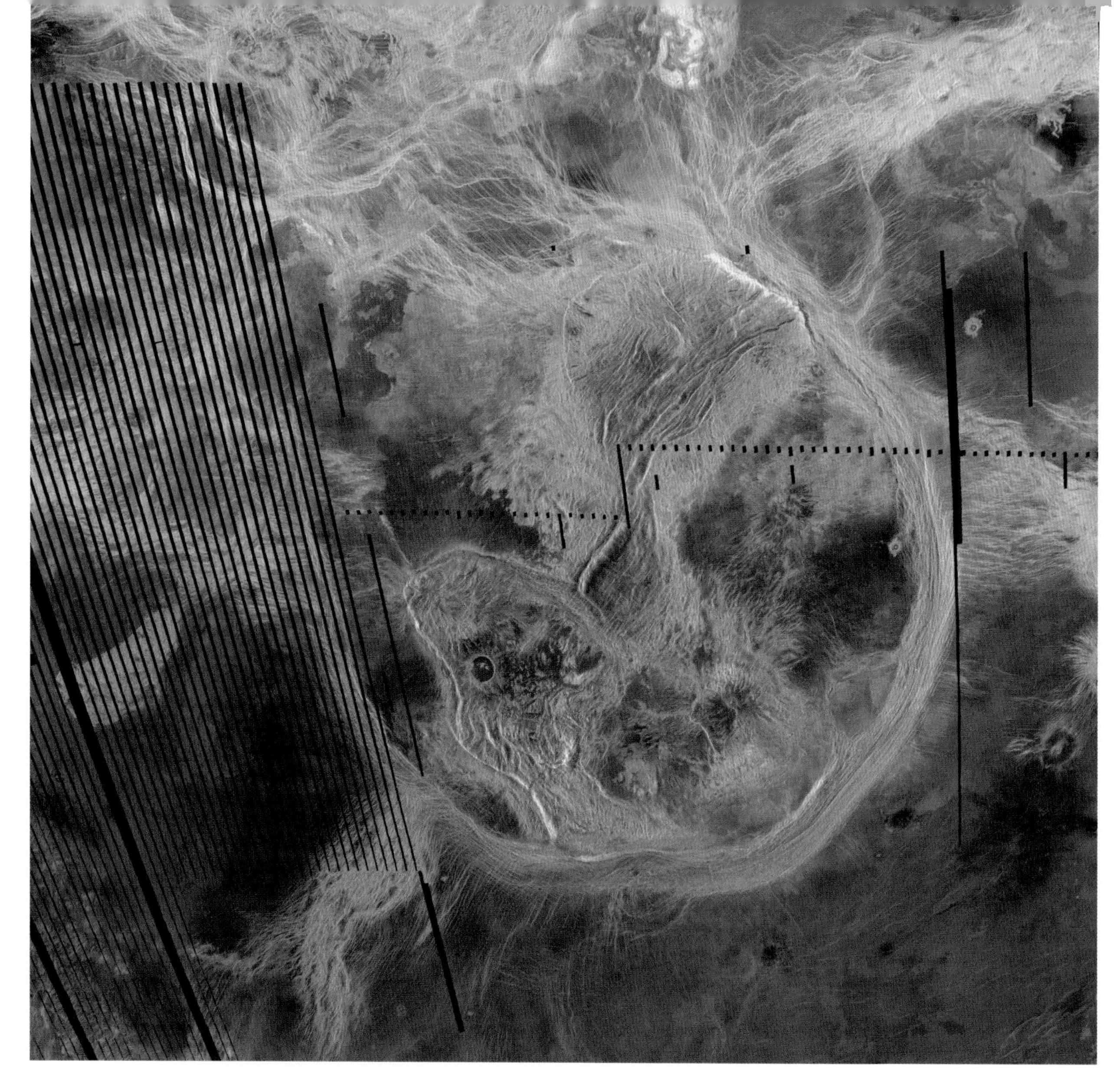

The complex ridges bounding a steep trough around Artemis Corona are about 120 km (74 mi) across and as deep as 2.5 km (1.5 mi). The dark stripes indicate areas where no data have been compiled.

several hundred kilometers across, Artemis's formidable ridges would stretch from Los Angeles to Seattle (about 1,827 km [1,135 mi]). Artemis is bordered by complex ridges bounding a steep trough. This encircling structure is about 120 km (75 mi) across, and is as deep as 2.5 km (1.5 mi).

Our first glimpse of coronas came from the Russian Venera 15 and 16 radar mappers. At first glance, they appeared to be the ghosts of immense impact basins. But it soon became evident that the Venusian surface was geologically far too young to have preserved the basins from impacts that ended around 3.9 billion years ago. Why are they so diverse, ranging from clear concentric fractures to faded disks? And what causes them? The diversity may be the result of age differences: coronas have been forming on Venus for a very long time and may be forming even today. And they may be caused by the upwelling of hot magma. As the magma pushes against the crust, the surface builds into a dome. The hot rock spreads out under its own weight and flattens. As the entire area cools, it sinks and cracks, forming a circular ring around a depression. Some of the old coronas are sunken almost beyond recognition. Newer ones have leaked trains of lava across the Venusian landscape. Coronas seem to be concentrated in certain areas of the Venusian globe. If the Venusian crust is of varying thickness, the coronas may be present where the crust is thinnest. Venus has no plate tectonics as the earth does, so the coronas may serve as a sort of pressure valve for the energy generated by Venus's shifting crust.

The coronas encircle some of the strangest territory on Venus. Although the planet is a lifeless wasteland, scientists have charted a veritable volcanic menagerie there, including ticks, anemones, and arachnoids. They've also identified figures that they've dubbed pancakes...and they're big ones. These whimsical names are assigned to formations that initially defy any geological—let alone logical—explanation. As they are Venusian, it is not surprising that they all appear to be related to volcanism.

Foundational to this collection of intriguing objects are the pancake domes. These eerie formations are roughly disk shaped and flat to slightly domed on top. They average 24 km (15 mi) in diameter and 762 m (2,500 ft) in height. Their steep sides are radar bright, indicative of rough surfaces, and the flat tops are often fractured. It is likely that the domes are the result of very thick lava flows. More than 150 have been identified on the planet. Earth has some similar formations known as domes (see chap. 2). But terrestrial domes are ten to one hundred times smaller than their Venusian counterparts.

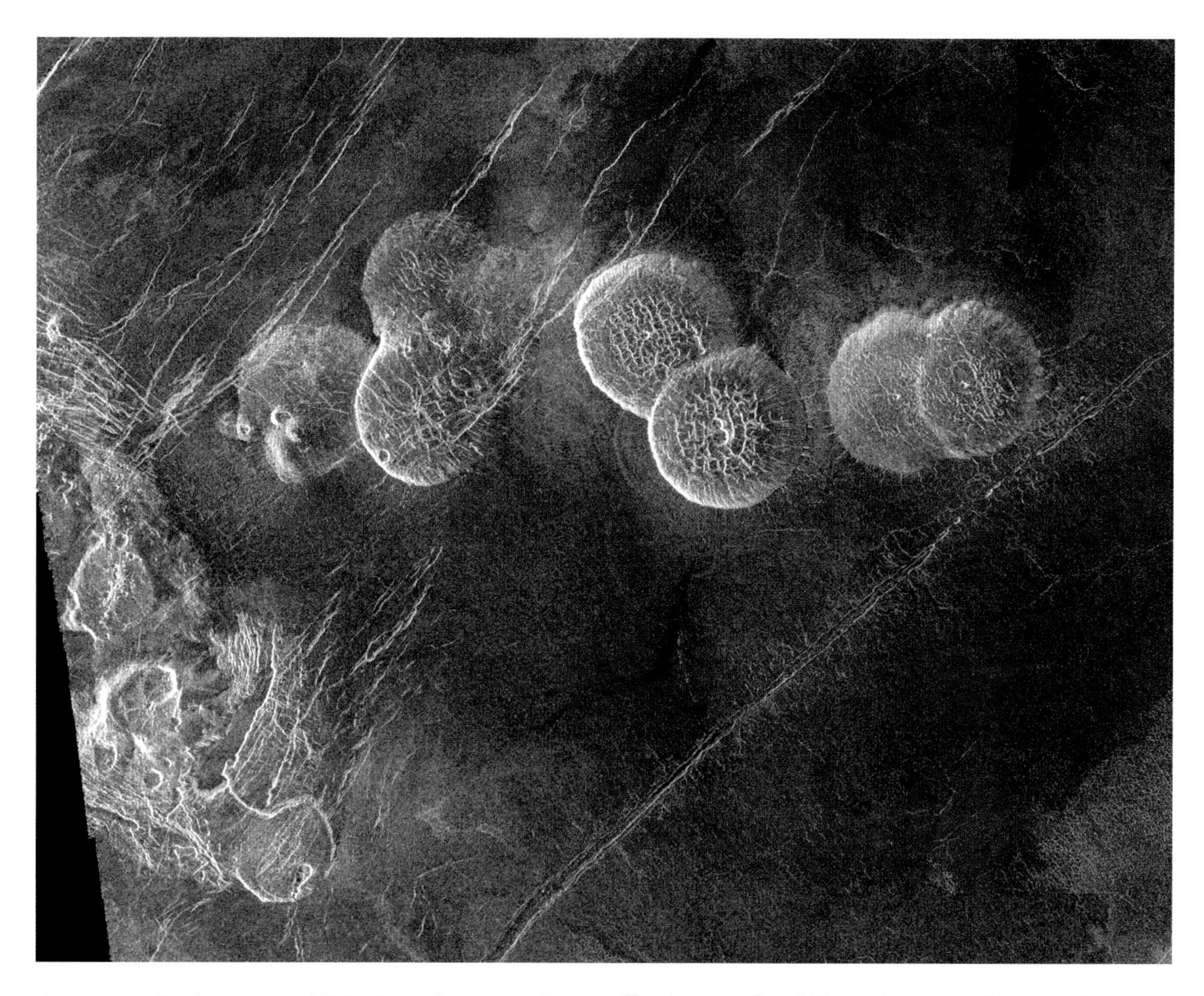

Seven pancake domes are visible along the border of Venus's Alpha Regio. The domes, which average 25 km (16 mi) in diameter, are thought to have been formed by thick, pasty lavas.

An excellent example of Venusian pancakes stands at the eastern edge of Alpha Regio, where seven domes loom above the plains. Complex fractures etch the top of each dome. These suggest that lava cooled on top, then was cracked as new lava expanded underneath—or drained away from inside—breaking the crust. Deep fractures also stretch across the plains below. Some of these troughs are covered by the domes, whereas others move across them, providing evidence of ongoing tectonic movement before, during, and after the domes themselves were formed.

Some of the pancake domes seem to give rise to other forms. One such volcanic beast is known as a "tick." Ticks, which appear to be degraded pancake domes, are usually associated with rift zones. Landslides sculpt ridges radiating outward from the dome itself. These ridges terminate in sharp ends that look like the legs of a tick. The legs may be scars of avalanches, or they may be the result of dikes running from the central body of the beast. At times, dark flows may exude from a summit caldera, flowing along lava channels. The summit tends to be concave. Approximately fifty have been identified so far. Ticks bear some resemblance to terrestrial seamounts, undersea volcanoes whose flanks have sloughed off into radial ridges.

Anemones are yet another member of the Venusian menagerie. These volcanoes exhibit lava flows arranged in overlapping petals extending outward in flowerlike patterns. The lava system usually occurs in association with fissure eruptions involving a series of elongated vents at the summit.

Rounding out the zoo are the arachnoids, volcanic domes surrounded by a cobweb of fractures and crests. Russian scientists first named these features after seeing vast concentric fractures spreading out from volcanic sources in Venera radar images. Their sizes range from 50 to 230 km (31 to 142 mi). Lines radiating beyond the arachnoids may be cracks or ridges resulting from upwelling magma that stretched the surface around them.

Many of the volcanic features appear geologically young, perhaps even active. One indicator of young age is found when studying craters on Venus. There aren't very many. With Venus's dense atmosphere, this makes sense. Many meteors burn up in the atmosphere on their way down or are so slowed by atmospheric friction that they leave a subtle scar upon impact. But the craters that remain are either in fairly pristine condition (newly made) or completely buried. There are not many in a state of decay somewhere in between. Some craters are partially covered in lava flows, but if volcanism were responsible for wiping out

The eruption of a tick volcano, a modified pancake dome, would be a spectacular event.

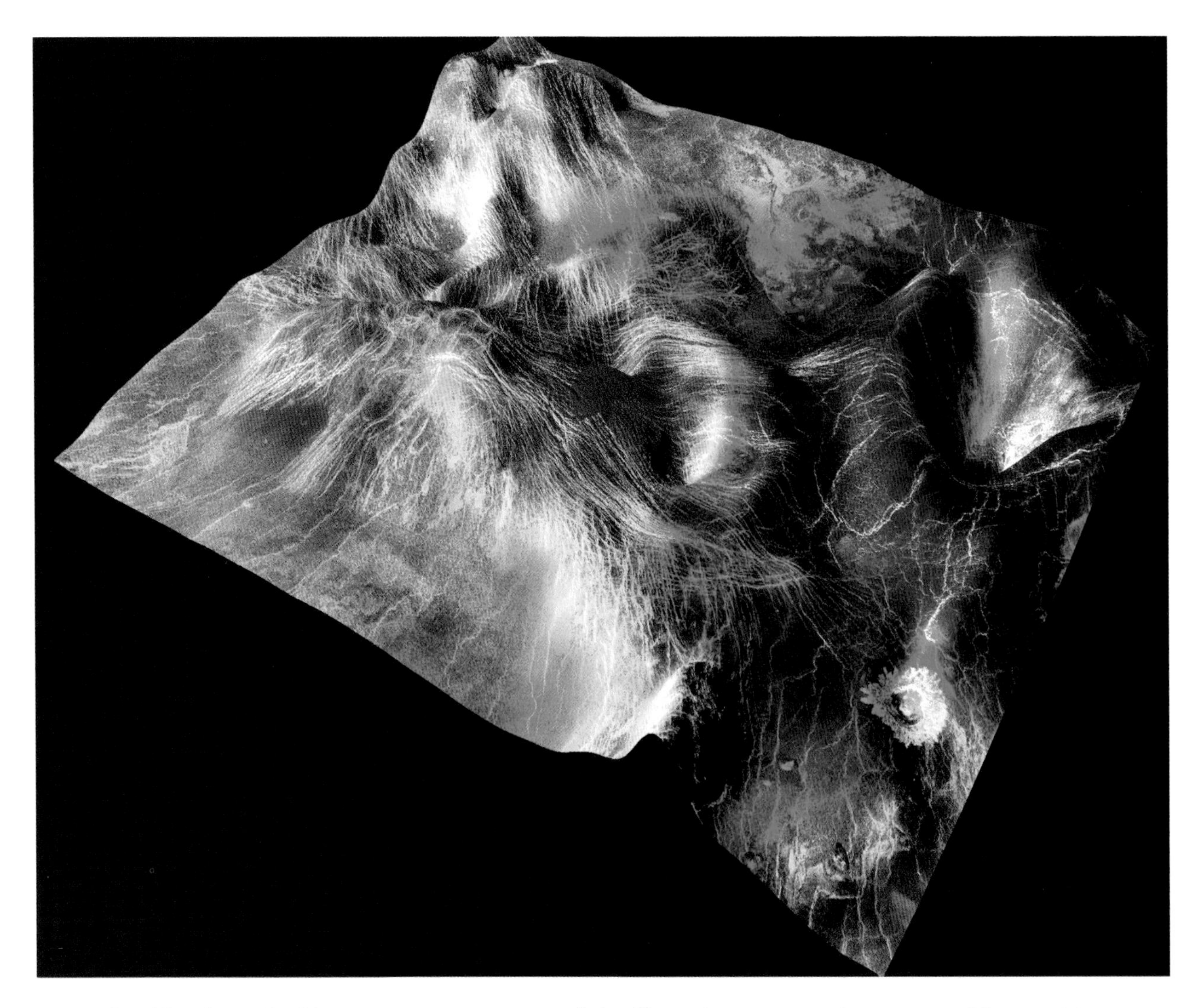

An arachnoid is shown in this computer-generated perspective view created from Magellan radar images; the vertical height is exaggerated. The arachnoid is about 100 km (62 mi) in diameter and 1 km (0.6 mi) deep.

most of the Venusian meteor sites we would expect to see many partially buried ones. This is not the case. Apparently something is obliterating the craters without a trace. Several theories have been put forth. One scenario suggests that the craters we see today are all recent impacts from a fairly limited, single event. This theory is not very satisfying; all evidence throughout the terrestrial planets points to the fact that there has *not* been a recent episode of cratering in the inner solar system.

Another proposal is that Venus had a moon that recently fell out of orbit, showering the planet with new craters. This is also an unsatisfying scenario; modern science abhors explanations involving special events or situations. The moon scenario is hard to disprove, but until other data come to light to support it, it seems a bit contrived.

A more convincing theory proposes that half a billion years ago, something triggered extensive volcanism on Venus after a period of relative quiescence. This planetwide spasm had the effect of resetting the crater counts on the

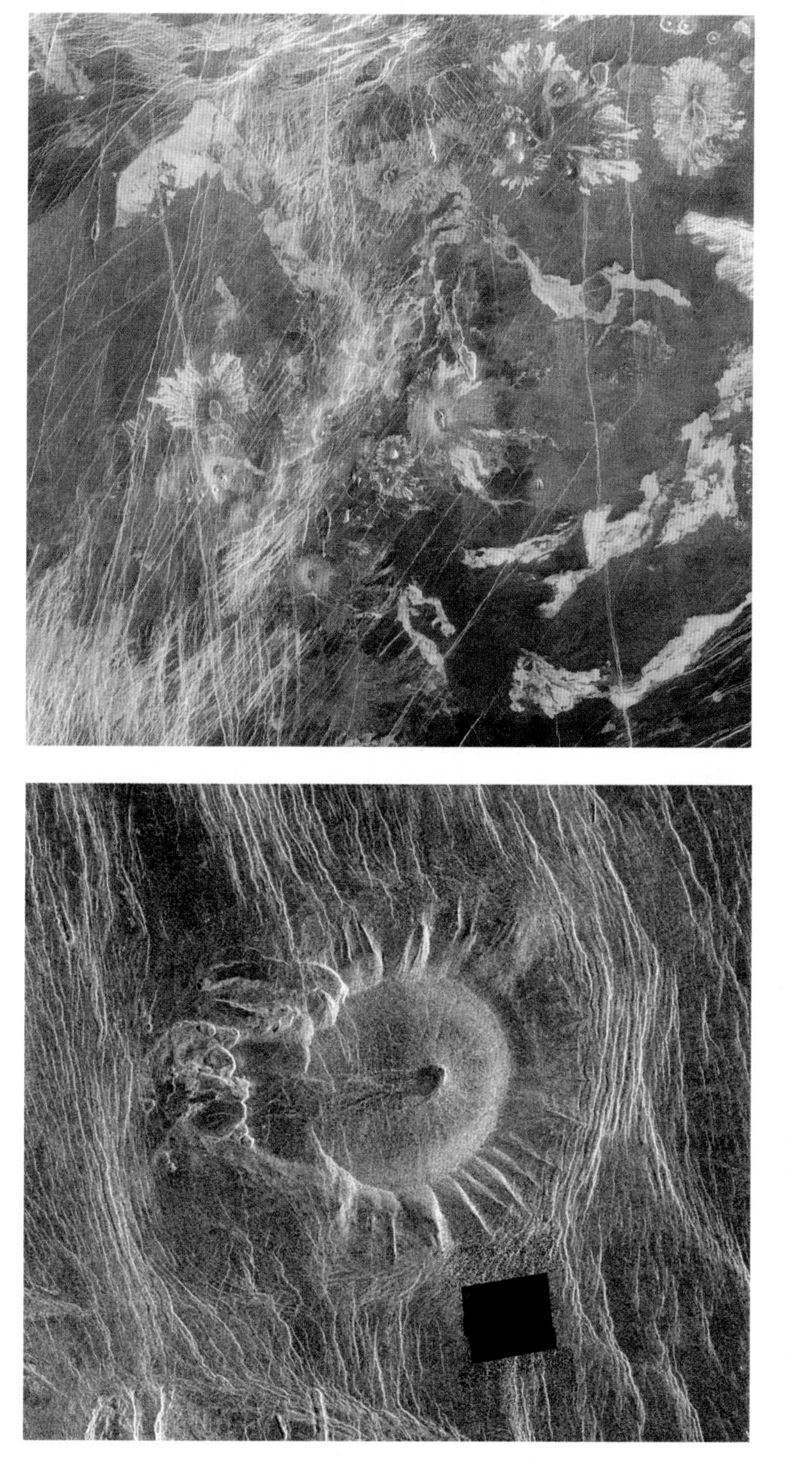

(*top*) Lava flows in the Atla region form flowerlike patterns called anemones. This Magellan radar image encompasses about 350 km (217 mi).
(*bottom*) Ticks are modified pancake domes. Note the landslides at the margins that give the appearance of legs. This Magellan radar image of the Eistla region encompasses 122 km (76 mi).

This three-dimensional perspective view of the Lavinia Planitia region of Venus shows three impact craters. In the foreground is Howe Crater, with a diameter of 37 km (23 mi). Danilova, with a diameter of 48 km (30 mi), appears above and to the left of Howe. Aglaonice, shown to the right of Danilova, has a diameter of 63 km (39 mi).

surface of Venus. From a 700-million-year-old surface, craters began to appear at the current, leisurely rate, leaving the numbers we see today. There is precedent for this on earth. At the end of the dinosaur age, massive volcanic eruptions centered on the Deccan Traps (in the region of modern India) belched so much ash and gas into the earth's atmosphere that the climate was significantly changed for an extended period of time. If the crust of Venus is as thick as it appears, it may be that Venus suffers from cataclysmic outgassing, involving hundreds or thousands of volcanic eruptions at a time.

Considering the well-preserved lava flows, crisp volcanic slopes, and modern-looking calderas, scientists were inspired to search for changes in volcanic structures during the Magellan mission. It was a difficult assignment: in most locations, Magellan overflew the same territory only twice, roughly an earth-year apart. No changes were visible, but in many instances, changes on terrestrial volcanoes are not evident from earth's orbit during a single year, so the lack of evidence is not definitive. But there is other evidence to support the idea of active Venusian volcanoes. In the late 1970s, Pioneer's ultraviolet spectrometer charted a steady decrease in the amount of SO_2 (sulfur dioxide) above the cloud tops. This decrease continued for several years. If the rate continued unabated, Venus would run out of sulfur fast, and we'd be able to set up super tanning booths for earth tourists. Many researchers, however, think that shortly before Pioneer arrived at Venus, a massive volcanic eruption sent billowing clouds of sulfur into Venus's upper atmosphere. It often happens on earth. Here, the sulfur combines with water and breaks down fairly quickly, but on Venus it would remain for some time. If this scenario is correct, then volcanic activity is present today on Venus. Later space missions may provide an unequivocal answer to the question of whether Venus's volcanoes are active, defunct, or merely dormant.

In this artist's rendering, primordial oceans of molten rock simmer on the plains of the early moon. At the horizon, the infant earth's landmasses consist of raised crater rims.

4 VOLCANOES ON DEAD WORLDS

If you look up at the sky on any moonlit evening, you can observe the ravages of planetwide volcanism. Across the face of our closest celestial companion, remnants of lava plains paint dark blotches into pictures as varied as those we see in the clouds. The features of the "man in the moon" were, at one time, oceans of lava. These dark plains are very ancient; the moon's most recent lava fountains probably fell silent more than a billion years ago.

Both the moon and the planet Mercury are small worlds. Small bodies cool more quickly than large ones like Earth, Mars, and Venus. The smaller worlds were not only less voluminous and less able to hold in heat, but also contained lesser amounts of radioactive material to stoke their internal fires. Today, both the moon and Mercury have cooled to the point where a thick crust caps any fluid magma that might simmer inside. Any molten rock today is buried forever beneath a thick, rigid crust of stone.

THE EARTH'S MOON

Even with the naked eye, it is easy to see that the moon has bright mountainous regions, called highlands or terrae, and vast expanses of dark materials, forming the lunar maria. These were once seas of lava. One can only imagine how spectacular they must have been, when long lava flows streaked across the surface.

The history of the moon, and of its volcanism, came into focus during the 1970s as Luna and Apollo surface samples began yielding their secrets to the world's planetary geolo-

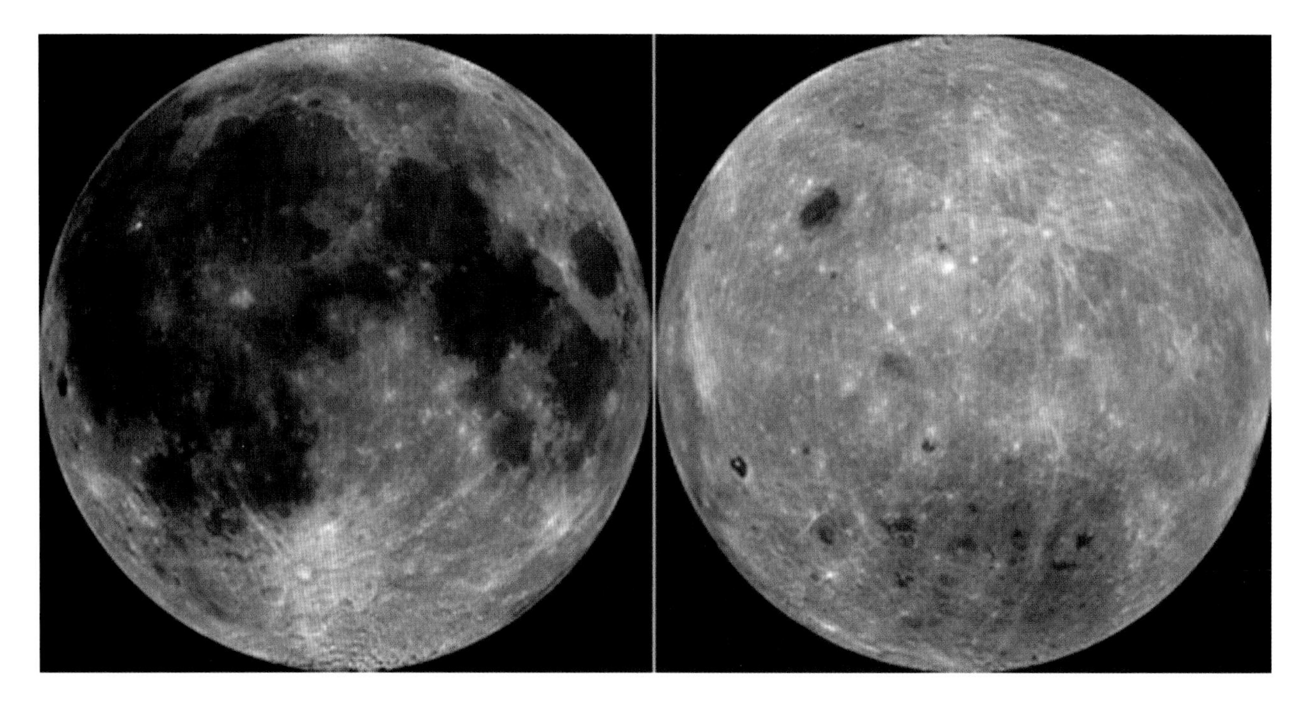

Both hemispheres of the moon were mapped in detail by the Clementine spacecraft.

gists. The twelve moon explorers from the Apollo missions brought back 380 kg (838 lb) of rock and soil, while Soviet Lunas returned 310 g (11 oz). Lunar samples fall into four categories. The first, which includes bright rock from the ancient, cratered highlands, represents the primitive crust of the moon that solidified nearly 4.5 billion years ago. This crust consisted of solid rocks resting atop a deep layer of melted magma. This liquid rock expanse is referred to as the magma ocean. For 600 million years, meteors and asteroids pummeled the crust into a cratered landscape that includes the highlands we see today.

The larger craters and immense impact basins were later filled by lava flows, some spreading more than 100 km (62 mi) in length. These basaltic lava flows represent the second class of lunar rock. Extensive lava flows probably began just after the heaviest rain of asteroids, about 3.9 billion years ago. (Earlier flows were largely obliterated by impacts, but may have started as far back as 4.3 billion years ago.) Radiometric dating shows that most flows occurred from 3 billion to 4 billion years ago. Their mineral makeup hints that their source was the moon's mantle itself, which partially melted at depths from 150 to more than 400 km (93 to 249 mi).

The other types of rocks on the moon are the lunar regolith—soil-fine material made from rocks pulverized by impacts—and the breccias, or broken and melted rocks formed by impacts.

Along with surface samples, mineral surveys taken from orbiting spacecraft like Clementine (1994), Lunar Prospector (1998–99) and the European Space Agency's Smart 1 (2003)

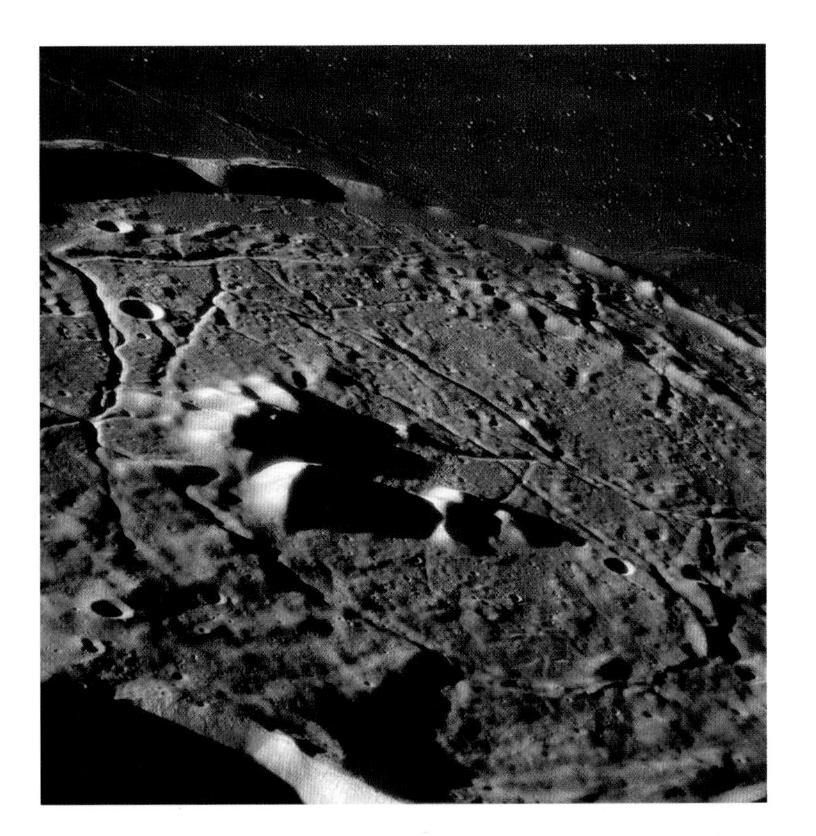

reveal the history of the moon written in stone. Early lavas contained high concentrations of titanium and may have had temperatures as high as 1400°C (2552°F). These lavas gave birth to Mare Tranquilitatis and Mare Serenitatus from 3.8 to 3.6 billion years ago. From 3.6 to just over 3 billion years ago, the moon saw its peak in volcanic activity. Lavas with lower titanium levels erupted to form Mare Imbrium and Oceanus Procellarum. Lavas of great mineral variety resurfaced vast areas of the moon, mostly on the near side.

(*top*) Ancient lava flows wash across the floor of the Gassendi Crater on earth's moon.
(*bottom*) Apollo 15 astronauts explored Hadley Rille, which is, on average, under 2 km (1 mi) wide and 400 m (1,312 ft) deep. The sinuous rilles are thought to be collapsed lava tubes or open channels along which lava flowed.

Apollo 15 astronauts took this photograph of Hadley Rille from the lunar surface.

Volcanism gradually declined, ending about 1 billion years ago.

The great volcanic age of the moon has left us with a diverse collection of volcanic remains over 16 percent of the lunar surface. The volcanic sources that have molded the earthly landscape are largely missing on the lunar surface. On earth, volcanoes usually form in chains, marking either the boundary of two plates or a hotspot under the moving crust. The moon was not influenced by plate tectonics. The sources (or vents) of most lunar lavas are not well preserved. Most lunar volcanic sources such as calderas, vents, and cones have been obliterated over time. Many of the flows left behind form circular deposits, pooled within craters and basins. In many locations, craters punched through the crust, enabling magma to flow to the surface. On the far side of the moon, where the crust is much thicker, evidence for volcanism is even scarcer.

The moon's family of volcanic formations includes

This image of Hadley Rille was taken from orbit by the Lunar Orbiter spacecraft. The width of the rille is about 1 km (0.6 mi).

lava flows, sinuous valleys called rilles, domes, and cinder cones. Most of the surfaces of lunar maria had their genesis as broad flows of runny lavas—lavas the consistency of cooking oil. Individual flows often appear to issue from arc-shaped depressions (rilles) near the edge of the basins where they pooled. Most of the vents are mercurial, hidden beneath younger flows or impact debris. In some cases, the original vents were covered as fluid lava flowed back into the vents at the end of the eruption, filling and masking the source.

The mare basalts are scored by sinuous rilles in many locations. Some of these valleys snake across the lunar landscape in discontinuous forms similar to collapsed lava tubes on earth. Other channels hundreds of meters wide continue across the desolate plains for hundreds of kilometers, with precipitous slopes dipping tens of meters down. Mare Imbrium's Hadley Rille, the V-shaped chasm explored by the Apollo 15 astronauts, is, on average, 1.6 km (1 mi) wide

Alphonsus Crater is seen in this image taken from orbit by Apollo 16 astronauts. Note the fractures on the crater floor, which is peppered with dark haloes that encircle rims of pits or craters.

and 400 m (1,312 ft) deep. The sinuous rilles are thought to be collapsed lava tubes or open channels along which lava flowed.

Although the most extensive type of volcanic deposits on the moon are the lava flows, implying that lunar eruptions were calm and flowing, there are some signs that explosive volcanism also occurred on the moon. Two types of pyroclastic (broken-up rock) features have been identified. In certain places, dark circular deposits drape the surface, some as large as 200 km (125 mi) across. These regions typically center at the edge of maria basins, suggesting that their source is the same as that of the lava flows. Spectral data show that the dark mantles are composed of fine volcanic particles in addition to the local rock distributions. Small, steep-sided cones and domes cluster on the maria plains. One such volcanic province spreads across the Maurius Hills region, where 250 domes, 60 cones, and more than 20 rilles are scattered across the dark flatlands. Still other eruption sources perch at the edge of lunar highlands. These large pyroclastic sites were probably generated by lava fountains with long-term flows, much like Strombolian eruptions on earth.

A second type of deposit darkens the floor of the ancient crater Alphonsus. The 90-km-wide (56-mi) crater floor is peppered with dark haloes that encircle rims of pits or craters. These craters, which lie along linear depressions, are thought to be sites of subdued volcanic eruptions. The haloes are a few kilometers in diameter. These eruptions were explosive. When magma stalled under the surface, gas

Most lunar glass was probably formed as a result of impact or volcanism. These samples were collected by Apollo astronauts.

bubbled toward the surface. As pressure built, the surface finally failed, explosively releasing gas, magma, and ash into the lunar vacuum. Debris fields associated with the dark haloes at Alphonsus were expelled up to 5 km (3 mi) from their volcanic sources. The moon's lower gravity and lack of atmospheric pressure contributed to the extent of the ejecta. The ejected material could have gone further had it not been for the fragmented surface of the Alphonsus floor. In some areas, dark mantle deposits along rilles extend up to 300 km (186 mi) across.

Lunar volcanism has yielded twenty varieties of glass in a rainbow of colors: purples, greens, blacks, and oranges. These glasses probably formed in violent lava fountains. Fine sprays of lava sailed through the lunar sky and cooled into tiny spheres as they fell back down to the surface. The composition of these colorful beads is similar to the most ancient and simple lunar materials. Lunar glasses give researchers clues to the composition of the interior of the ancient moon during its volcanic epoch.

Lunar eruptions occurred over a long period of time, starting early in lunar history, perhaps as early as 4.3 billion years ago, and ending about 1 billion years ago. By comparison, the volcanoes we see on earth are young. Eruptions on the moon raged long before life existed on the earth.

MERCURY

At about the time the moon was geologically dying, the planet Mercury was probably nearing its last throes of volcanic activity. Although impact scars camouflage Mercury's

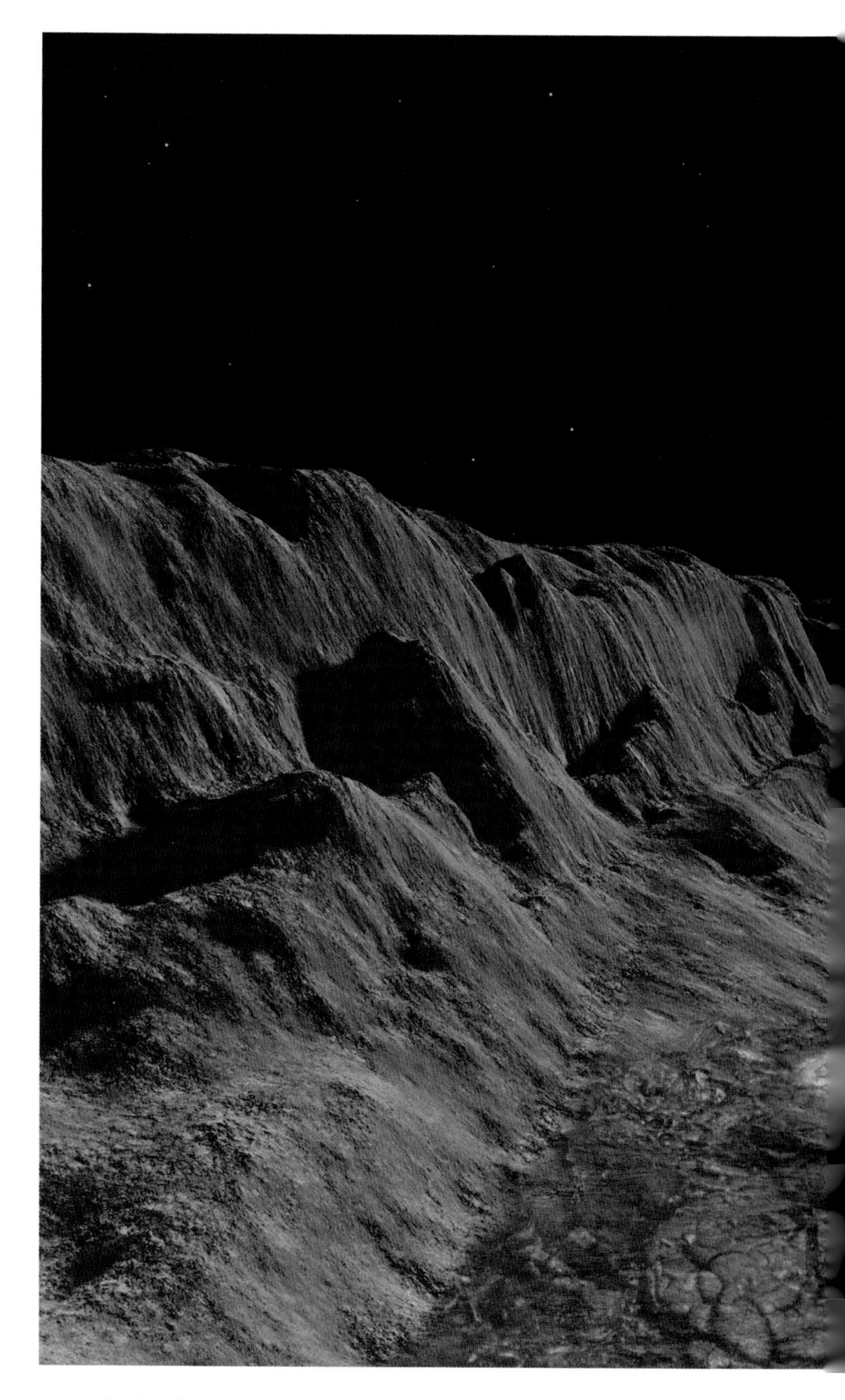

The sun rises over the desiccated surface of Mercury in this artist's rendering. Volcanic vents and lava flows can be seen on the plains leading to a dramatic scarp at left.

record of volcanism, cratered surfaces actually help us establish timetables for Mercury's volcanic age. The small planet is devoid of craters less than 50 km (31 mi) across, leading scientists to conclude that anything smaller was buried under lava billions of years ago. But the search is on for specific volcanic forms on Mercury.

Mercury is the poster child for gleaning new insights from old data. The last close reconnaissance of Mercury was carried out on March 16, 1975, when, in the last of

three flybys, the Mariner 10 spacecraft imaged 45 percent of the surface of the smallest terrestrial planet. The Mariner 10 data were recently recalibrated so that compositional differences could be teased from the decades-old data. Mark Robinson, then at the U.S. Geological Survey in Flagstaff, and Paul Lucey of the University of Hawaii, corrected the old data, evening out variations across the pixel array of the spacecraft using master images taken before launch in a laboratory at the Jet Propulsion Laboratory. Using these

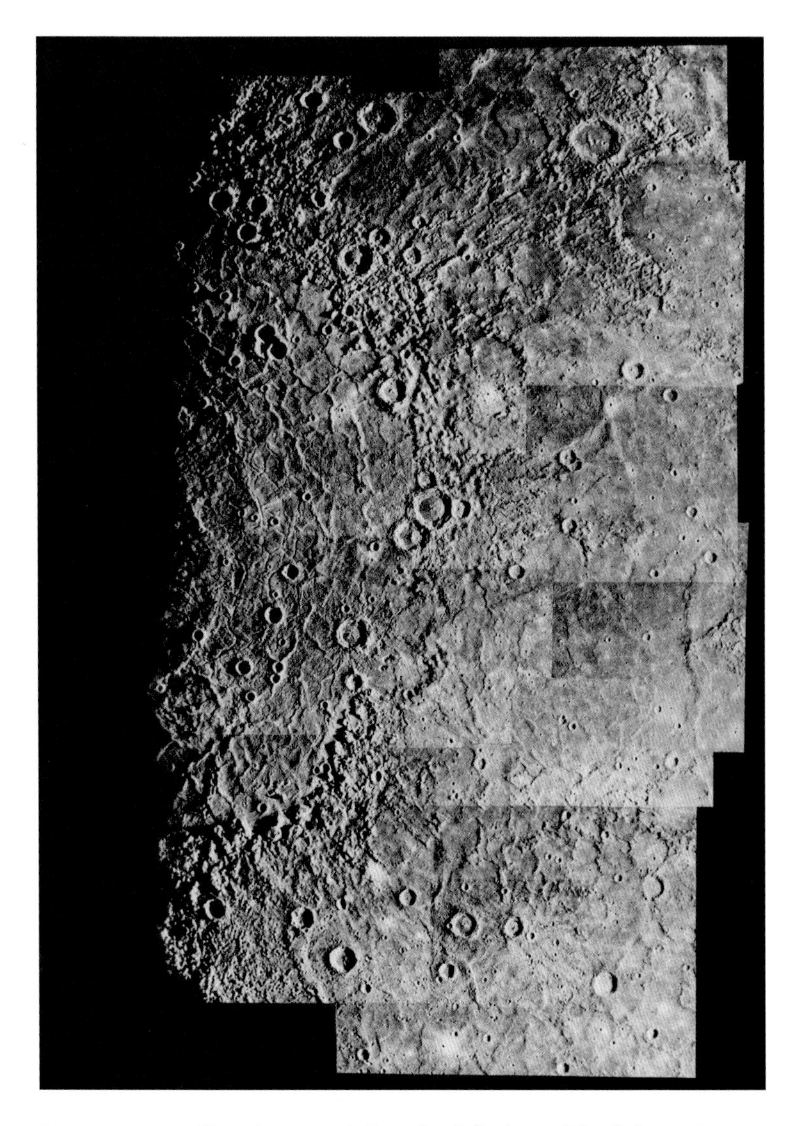

Rilles snake across the Mercurian landscape in these images from Mariner 10. Many of these rilles have sinuous forms reminiscent of those on earth's moon, leading some researchers to think the rilles are remnants of lava channels or collapsed lava tubes.

images, as well as images taken by Mariner 10 of the relatively featureless clouds of Venus, the researchers removed blemishes from the data set. They then used spectral studies—studies of different kinds of light—from lunar samples and telescopic images of lunar spectra to compare with the imagery of Mercury. Advances in spectra studies of the moon have given researchers a better handle on how various minerals and glasses respond to reflected light. Robinson and Lucey combined Mariner 10 data taken in two wavelengths, creating false color images of the Mercurian surface. The resulting colors revealed variations in the composition of the surface.

The new research presents a picture of a more volcanically active planet than scientists initially thought. One telltale sign that the researchers looked for was a specific type of difference in composition from one area to another. For example, the smooth plains west of the crater Rudaki seem to pool at their boundaries, forming embayments at the edge of higher ground. While the plains are similar

MOLTEN MERCURY

Researchers have long debated whether Mercury's core is molten or whether it is frozen and geologically dead. Mercury's internal status has a direct bearing on the history of that planet's volcanism. If the core is still molten, volcanism may have been present on the surface for a longer period than if the core is cold. The Mariner 10 spacecraft revealed a magnetic field just 1 percent of that on Earth, leading many to think that Mercury's core is no longer hot. (Planetary magnetic fields are linked to molten cores.)

Using the Arecibo radar dish in Puerto Rico, scientists were able to detect slight variations in Mercury's spin. The scientists had only a 20-second window each day to bounce radar off the surface of the planet. These studies show shifts in the planet's rotation, indicating that Mercury's core is still at least partially molten. The results may indicate the presence of sulfur in Mercury's interior, which would lower the melting point of Mercury's iron core.

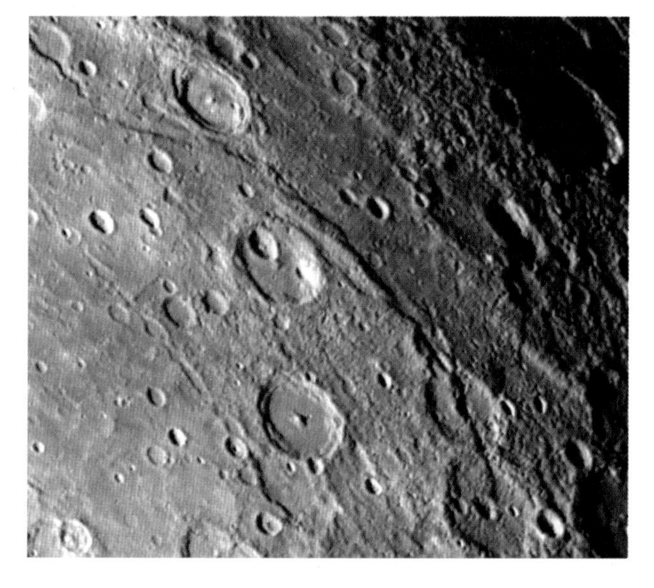

in form to lava plains on Venus or the moon, earlier studies lacked enough information to determine whether the flows differed from nearby materials. With the new data in hand, Robinson and other researchers have shown that these embayed plains are, in fact, distinct from their surroundings. This is consistent with the suggestion that the plains are volcanic in origin, and constitute flows that have covered areas made up of different material. While some researchers proposed that these plains were formed by

(*left*) Mariner 10 obtained this image of a 65-km (40 mi) crater and the scarp crossing its floor. The scarp may be the result of thrust faulting.
(*right*) This image of a 300-km-long (186-mi) scarp on Mercury was captured by Mariner 10.

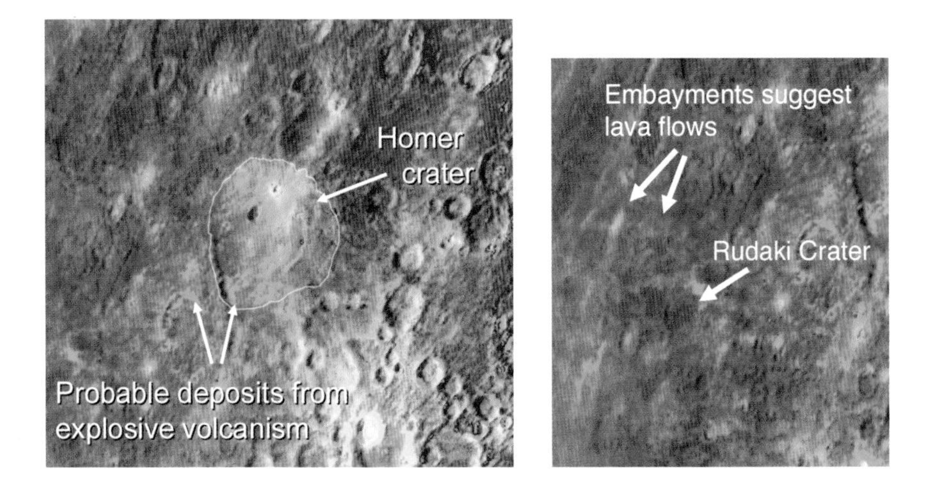

(*left*) The rim of Homer Crater may support evidence of volcanic deposits. (*right*) Lava flows may have embayed the plains adjacent to Mercury's Rudaki Crater.

slurries of material from impacts, the smooth plains contain fewer craters than ejecta blankets surrounding craters. This indicates a younger age for the flows than the cratering epoch of our early solar system.

Mariner 10's resolution was inadequate to discern the kind of small vents, domes, and cones that we have seen on the moon. Large shields are nonexistent in the imaged areas of Mercury. But in addition to the embayed plains, whose borders are just within the resolution of the best Mariner 10 images, rilles snake across the Mercurian landscape. A web of these steep-sided valleys scores the smooth floor of the great Caloris Basin, some 1,300 km (808 mi) across. Many of these rilles, which have sinuous forms reminiscent of those on the moon, may be remnants of lava channels or collapsed lava tubes. As the rilles are associated with Mercury's plains, they add to the evidence that the smooth areas are volcanic in origin.

Some Mercury terrain may also betray signs of pyroclastic eruptions. Two darkened sites adjacent to the Homer Crater cover a roughly circular footprint. Their borders fade into the surrounding terrain, much as one would expect of pyroclastic deposits. Although they bear superficial resemblance to ejecta blankets, no crater can be found at their centers. Both sites lie along a linear section on Homer's degraded rim. This segment of the crater's border resembles some volcanic fissures on the moon that lie along the rims of impact basins.

Mercury is certainly large enough to have carried on an extended period of volcanism, both from the heat of

In this artist's rendering, the MESSENGER spacecraft prepares for orbital insertion around Mercury.

accretion and from radiogenic heating in its core. Volcanic eruptions exploded from the earth's moon for at least 2 billion years. Mercury is a third larger than the moon, so its volcanism may well have lasted longer. However, its crust may be thicker than lunar crust, so that volcanism may have been subdued.

Definitive answers to the mysteries of Mercury's volcanic history will have to wait for closer scrutiny of the planet. The *ME*rcury *S*urface, *S*pace *EN*vironment, *GE*ochemistry, and *R*anging (MESSENGER) spacecraft is en route at this writing and will carry out a series of flybys beginning in January 2008. The flybys will culminate in orbital insertion in 2011. From orbit, MESSENGER may well unlock the door to Mercury's past. If Mercury follows the tradition of its planetary siblings, it holds many surprises for us.

An artist imagines a volcanic eruption on Io, the innermost of Jupiter's four largest moons. A fissure in Io's Tvashtar caldera spouts a fire fountain of molten rock some 2 km (1 mi) into the sky as lava flows flood the caldera floor.

5 A VISIT TO DANTE'S INFERNO

Imagine a world of scorching lava flows juxtaposed against frozen wastelands, a world tormented by volcanic upheaval, twisted by powerful gravity and radiation. In his *Inferno,* the fourteenth-century poet Dante Alighieri envisioned Hell as a world of belching fire, billowing smoke, sulfurous fumes, and tortured landscapes, with a great frozen sea at the center.

Dante's vision of Hell is remarkably similar to a real world 63 million km (39 million mi) from earth. This Dante-esque panorama is Io, the innermost of Jupiter's four largest moons. There, lava gushes like searing oil across multicolored plains. Geyserlike plumes rocket 500 km (310 mi) into the airless sky, raining down a hail of frozen sulfur. Mountains thrust out of the ochre plains like icebergs from an arctic ocean, towering above red, yellow, and white blankets of sulfur. Fountains of lava explode from cracks several kilometers long. Snowy sulfur dioxide frosts rocks and valleys, contrasting with strange greenish areas that planetary scientists have nicknamed "golf courses." The volcanoes that sculpt the Ionian landscape are some of the most alien in our solar system. Io is a strange realm indeed; one that continues to challenge the best scientific minds of our time.

Galileo, who discovered Io, could not have anticipated all we know now about Io's tumultuous character. Nor could he guess that this little point of light would cause such upheavals in his own life. On a cool January evening in 1610, he turned his new telescope toward Jupiter for the first time. His optical contraption was based on the work of

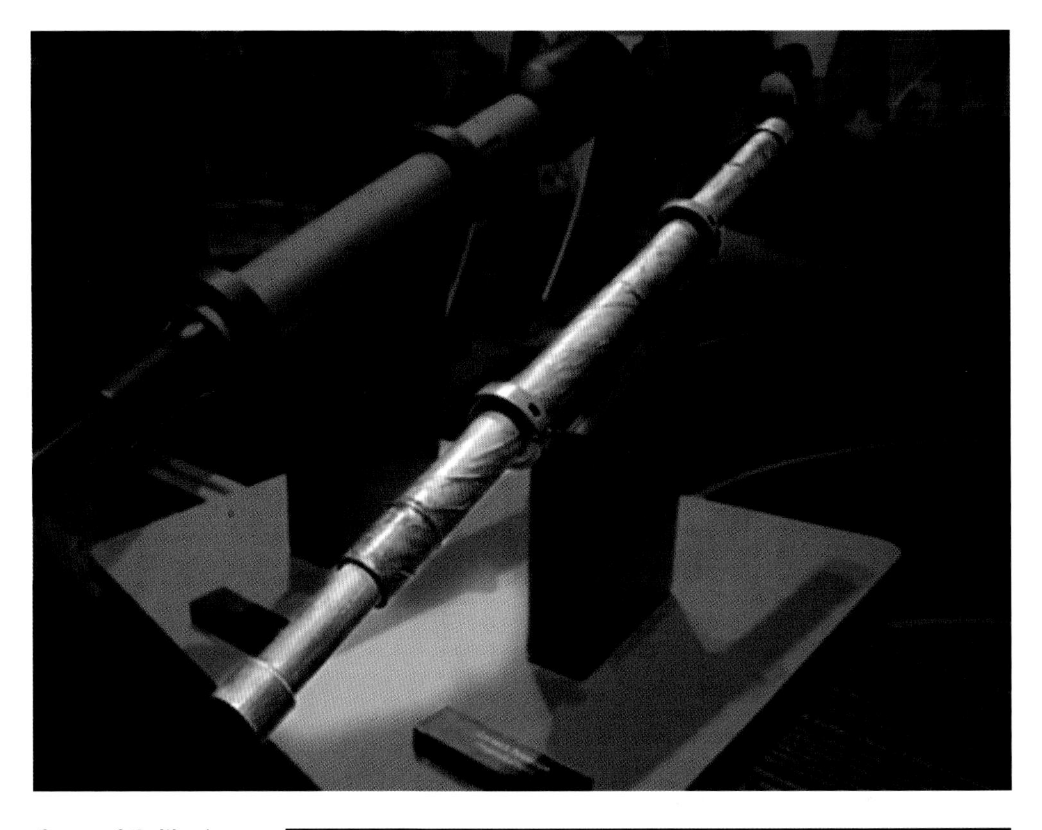

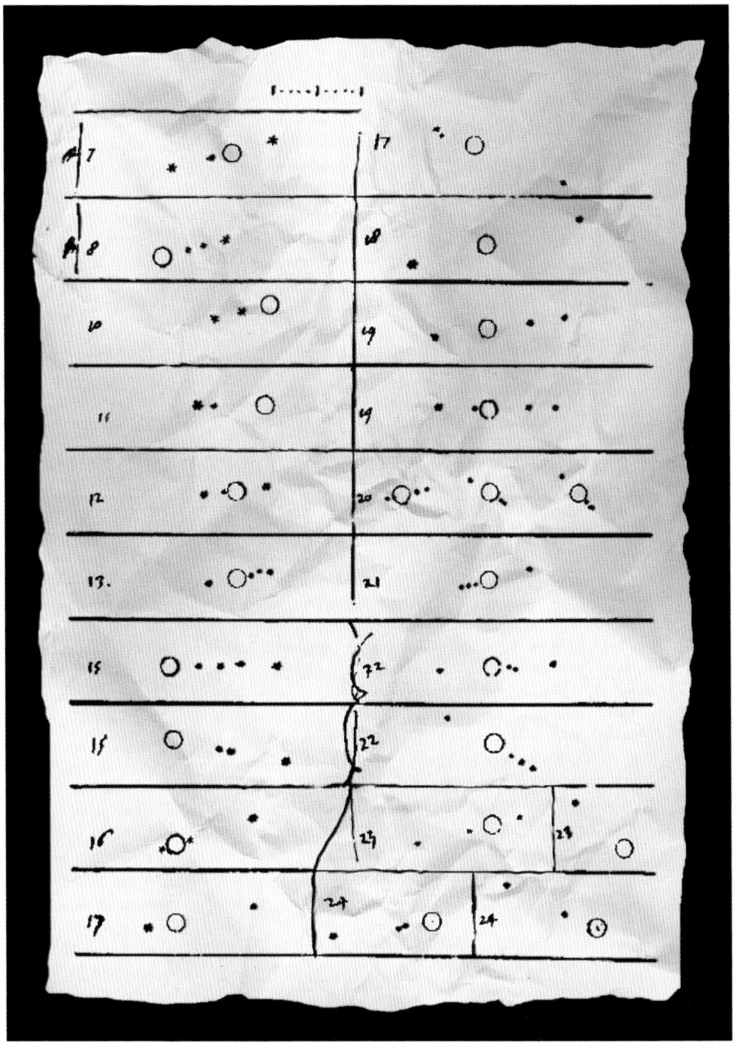

(*top*) A reconstruction of one of Galileo's telescopes is on display at the Imiloa Planetarium, Hilo, Hawaii.
(*bottom*) Galileo charted the movements of starlike objects traveling back and forth across the face of Jupiter. He named the moons of Jupiter "Medicean stars" as a "gift" to Grand Duke Cosimo II de' Medici.

the Dutch optician Anton van Leeuwenhoek. Others had experimented with small spyglasses, looking at ships across the bay or neighbors on their balconies doing who knows what. But with his improved 30-power instrument, Galileo decided to look up to the heavens. What he saw there changed his life, and set him on a dangerous path.

Night after night, Galileo charted the movements of starlike objects traveling back and forth across the face of Jupiter. On the night of January 7, three stars glowed near the orb of Jupiter, two to the east and one to the west, all in a perfectly straight line. The next night, all three were to the west, still in a nice, neat line. In subsequent nights a fourth appeared, following the same straight path. Galileo came to the uncomfortable conclusion that these "Medicean stars"—as he first called them—were circling Jupiter. The revelation was uncomfortable because of accepted scientific theory at the time. The Greek philosopher Aristotle (384–322 BC) asserted that all celestial objects—including stars—revolved around the earth, which was the center of the universe. This position was adopted later by the Roman Catholic Church, which held great power in postmedieval Europe. Galileo's discovery implied otherwise, reinforcing recent work by Nicolas Copernicus, who placed the sun, not the earth, at the center of things.

Galileo had many friends within the church, associates who shared his view that science was simply a means of figuring out how the great creator of the universe went about designing things. As Galileo put it, "Philosophy is written in this grand book—I mean the universe—which stands continually open to our gaze, but it cannot be understood unless one first learns to comprehend the language and interpret the characters in which it is written." Galileo's attempt to "comprehend the language" of nature cost him, at various times, his salary, his good name, and his personal freedom. Nevertheless, he was eventually proved right. Those troublesome little stars shuttling back and forth across the face of Jupiter were worlds in their own right, the largest moons of the largest planet in our solar system. The innermost of them, Io, would send waves of change across the scientific community.

With the advent of modern telescopes, researchers began to suspect that Io was different from other moons. Jupiter's family of known moons consisted of Amalthea, an irregular-shaped rock several hundred miles across, and the four giant moons discovered by Galileo—Ganymede, Callisto, Io, and Europa—now known as the Galilean satellites. The signature of the light coming through telescopes told researchers that the surfaces of the Galilean moons were covered by water ice. All, that is, except Io. Io was as bright as its icy

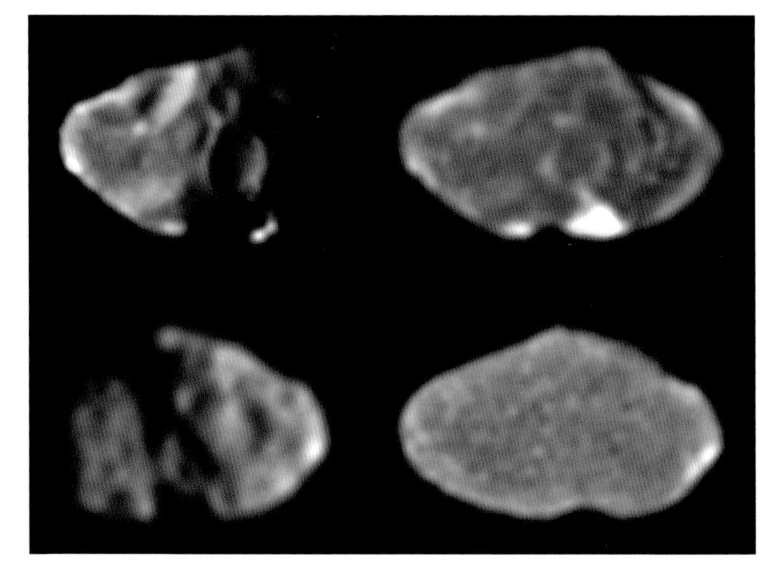

(*top*) Jupiter's four largest moons, known as the Galilean satellites, are shown to scale. From left to right they are Io, Europa, Ganymede, and Callisto.
(*bottom*) Amalthea is one of a host of moons circling Jupiter. These four views show the irregular shape of Amalthea.

siblings, but its spectrum was as dry as dust. Astronomers detected strange elements like sulfur and sodium. Why was Io so fundamentally different from the other Galileans?

Armed with ever-stronger telescopes, scientists began assembling a picture of Io. With a diameter of 3,636 km (2,259 mi), Io is about the size of earth's moon. Io and the other Galileans are tidally locked in their orbits, meaning that the same hemisphere always faces toward Jupiter. Io orbits Jupiter at a distance of 421,800 km (262,100 mi), circling deep within the planet's deadly radiation fields. Its day, equivalent to its orbital period, is forty-three hours. And unlike the gray white moons, Io appeared to many observers as a brilliant orange. Nor does Io have the same frigid surface as the other three Galileans. Instead, observers detected a hundred times as much heat flowing from its surface as flows from earth. These results were hard to credit.

Another mysterious feature of Io was a great, doughnut-shaped cloud of electrically charged sodium atoms that seemed to envelope its orbit around Jupiter. This "Io

PREDICTIONS OF A COSMIC TUG-OF-WAR

Three researchers—Stanton Peale, Patrick Cassen, and Raymond Reynolds—realized that Io might be a dynamic little world, constantly sculpted by the push and pull of gravity from Jupiter and the other massive Galilean satellites. Such a gravitational taffy pull is called tidal heating. Peale's team proposed that tidal heating might be powerful enough to generate volcanism that would obliterate craters and create geologically young features. Their paper, which was published just days before Voyager 1 reached Jupiter, turned out to be prescient. Tidal heating is responsible for making Io the most volcanically active world yet found in our solar system.

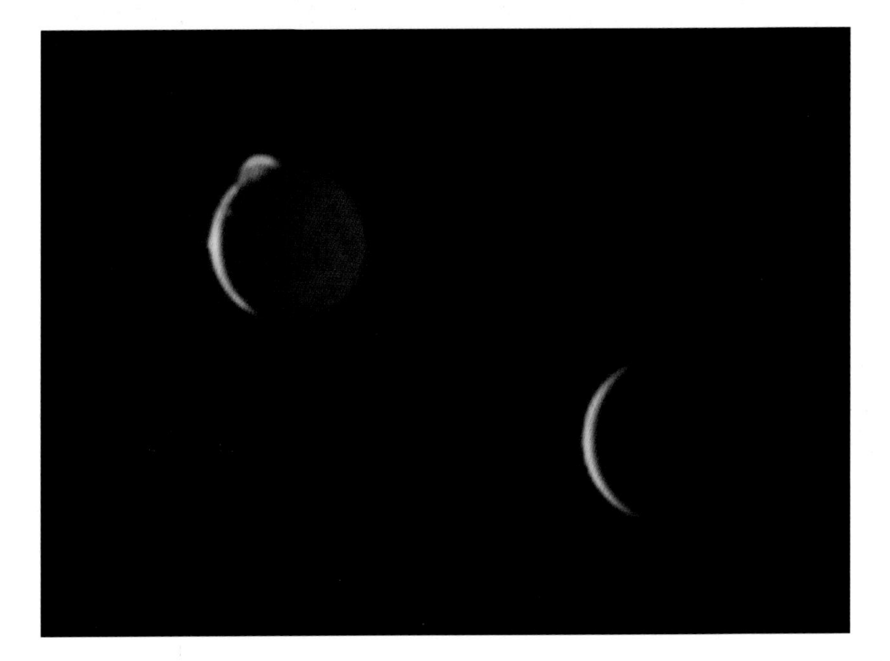

Volcanic plumes on Io can spew material 300 km (190 mi) into space. This image of Io (*left*) and Europa was taken by NASA's New Horizons spacecraft on March 2, 2007.

torus" was somehow linked to the massive magnetic fields sweeping around Jupiter. Io's location around Jupiter also appeared to govern powerful radio emissions. To understand this complex moon, closer inspection was called for.

Scientists awaited the first close-up views of Io with great anticipation. Voyager 1 was due to swoop within 17,000 km (10,600 mi) of the mystery moon on March 5, 1979. Several years before, Pioneer 11 had given us fuzzy, distant views. Scientists hoped that Voyager's comparatively advanced imagers could help explain the mysteries of Io.

Many in the scientific community expected Io to be a crater-pocked, geologically dead moon much like the earth's own. Others thought Io might have a geologically exciting surface influenced by Jupiter's mighty gravity.

Three days after its closest approach to Io, on March 8, the Voyager spacecraft took a series of overexposed images of stars for navigation purposes. One of these images caught the limb (sunlit edge) of a crescent Io. Hovering 300 km (190 mi) above the surface of the moon was an

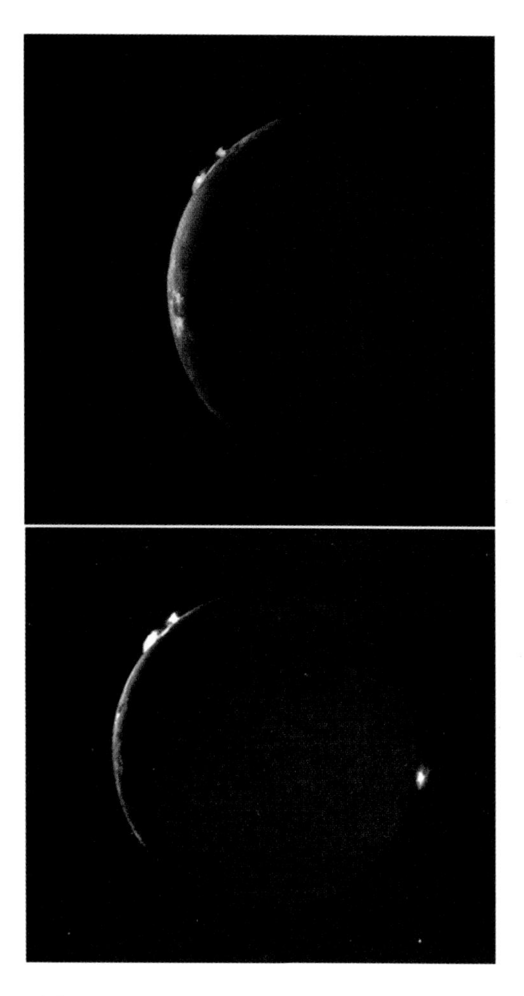

(*left*) The Voyager 2 "volcano watch" recorded plume eruptions on Io over several days.
(*right*) This false-color image of the Loki plume taken by Voyager 1 shows variations in particle size and in the amount of particles in different parts of the plume. The part of the plume that is brighter in ultraviolet light (blue in this image) contains more fine particles and is much more extensive than the denser, bright yellow region near the center of the eruption.

umbrella-shaped plume. A similar feature was visible at the terminator (the boundary between day and night). Navigator Linda Morabito recognized the ghostly apparition as a cloud. Knowing that Io had no atmosphere, the only logical conclusion left to her was that an incredibly violent volcanic eruption had blasted the umbrella of gas and dust into space. Researchers on the imaging team agreed, and infrared sensors confirmed hotspots across the face of the little moon. We have since discovered that the push and pull of Jupiter and the other Galilean satellites causes Io's ground to rise and fall some 46 m (150 ft) each day. This flexing heats the interior of the little moon. All the energy must find release somewhere. It comes out in the form of the moon's omnipresent volcanoes. Io's ongoing inferno, driven by tidal heating, helped explain the existence of the torus and Io's torrential flow of heat. The search was on for active volcanoes.

In all, Voyager 1 found eight active volcanoes, along with hundreds of calderas and vast deposits of lava and sul-

THE DISCOVERY

In this image taken by Voyager 1, the Pele plume is seen on the limb (the sunlit edge of the planet); on the terminator (the boundary between day and night) the top of the Loki plume is illuminated by sunlight. When navigators commanded Voyager 1 to take this overexposed shot of Io to search for background stars, the image showed wisps of a mysterious material floating over the moon. Linda Morabito, the flight navigator, was the first to recognize that the light areas were actually features associated with Io. But what could they be? Io has no atmosphere, so the drifting fog couldn't be clouds. Something was coming out of Io.

Voyager imaging team members rushed to their computers to search for cloudlike structures in other Voyager photos. Several were found, and many were the shape of a volcanic eruption in a vacuum. Voyager—and Linda Morabito—had discovered live alien volcanoes.

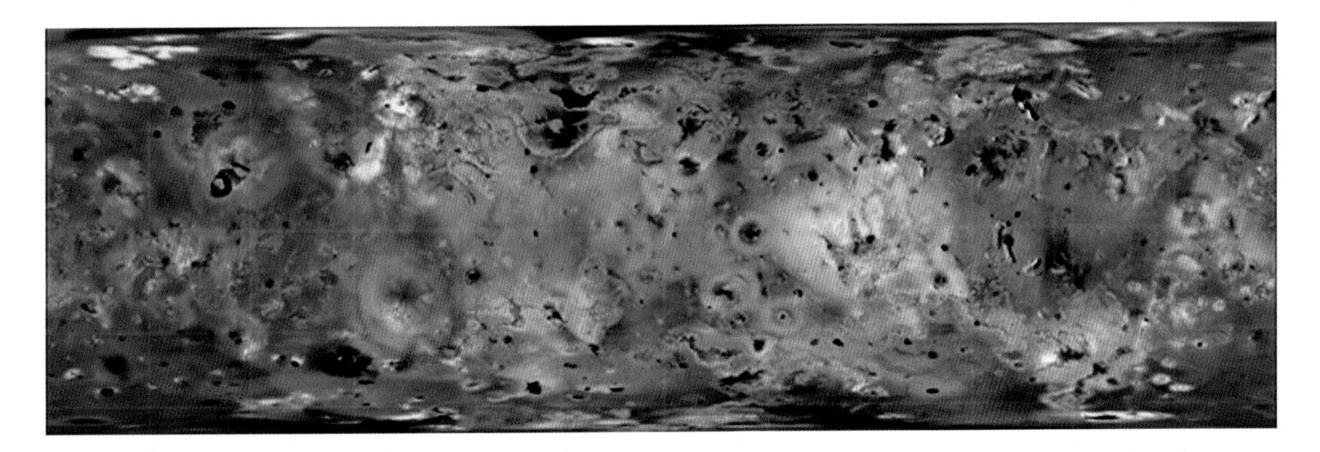

Deposits of sulfur and sulfur dioxide color Io's surface. Red deposits indicate recent plume activity; black areas represent flowing lavas.

fur. Four months later, a reprogrammed Voyager 2 embarked on a ten-hour "volcano watch." The assembled images captured graceful volcanic umbrellas of ballistic debris and ragged plumes erupting hundreds of kilometers into space.

The twin Voyagers gave us telling glimpses of Jupiter and its satellites, but nearly twenty years elapsed before we got a closer, in-depth look. In 1995 the massive Galileo spacecraft settled into orbit around Jupiter for a long study. Its eight-year tour of the Jovian system has changed our fundamental understanding of Io and its siblings.

Before Galileo arrived, scientists wondered if the craft would detect any new hotspots on Io. In fact, the Voyagers, Galileo, and ground-based studies have now found more than 150 volcanic sources. Scientists estimate that there may be 300 to 500 across the moon's tortured face. This number is remarkable, considering that the surface area on Io is only equivalent to the North and South American continents combined. The moon's surface area may contain nearly as many volcanoes as the entire land area on earth, where there are currently 500 to 600 active volcanoes.

Io's surface is shattered and molded by three main types of volcanic activity: lava lakes, lava flows, and violent eruptions. The first type (and the most common) is the lava lake. These collapsed holes punch through the plains, leaving boiling cauldrons of lava. The classic example of these formations is a sulfurous basin called Loki (named after the Norse God of fire). Loki is a volcanic crater, or caldera, filled with a lava lake some 200 km (124 mi) across. Within its dark, viscous liquid is a gigantic islandlike feature covered by frozen sulfur dioxide. The island is about the size of Rhode Island. (It is most likely anchored and made of silicates, not floating.) Galileo infrared observations show an incandescent shore along one edge of the lake, similar to the Kilauea and Halemaumau lava lakes on Hawaii. In these lakes, cooled lava forms a crust that breaks up as it collides with the caldera wall, exposing a spider web of glowing lava cracks.

The second type of volcanic activity on Io is the eruption of lava flows. Molten rock and possibly sulfur have raced across nearly every square meter of the hellish moon. The longest flow is called Amirani, which stretches across 300 km (190 mi) of rolling plains. Many flows are insulated, meaning that a crust of cooled lava covers them. Magma frequently breaks out of the crust many kilometers downstream. An archetypal flow is associated with the volcano Prometheus, whose eruption plume reaches 100 km (62 mi). In the seventeen years between the Voyager and Galileo flights, the Prometheus plume moved some 70 km

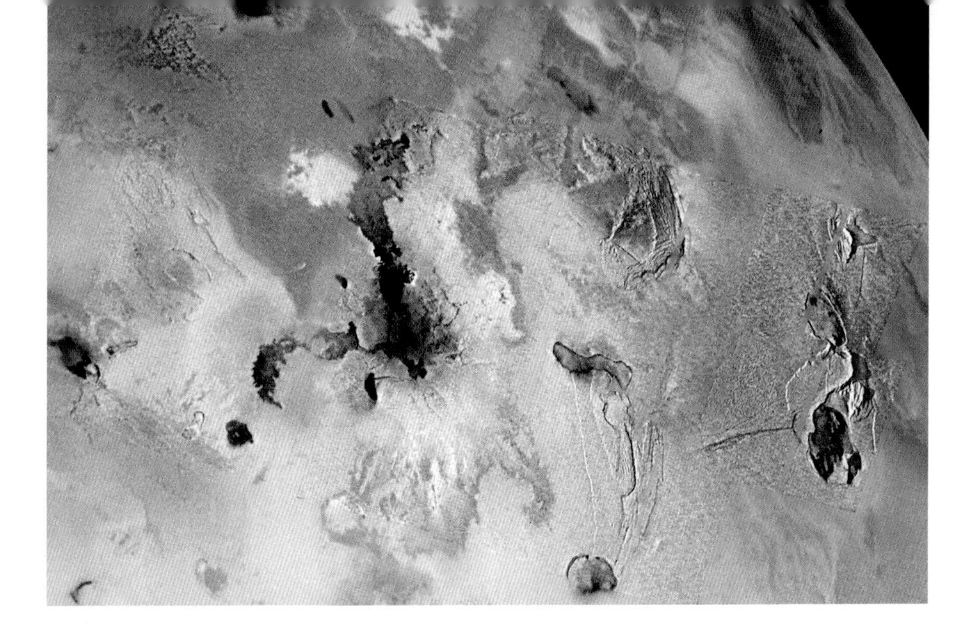

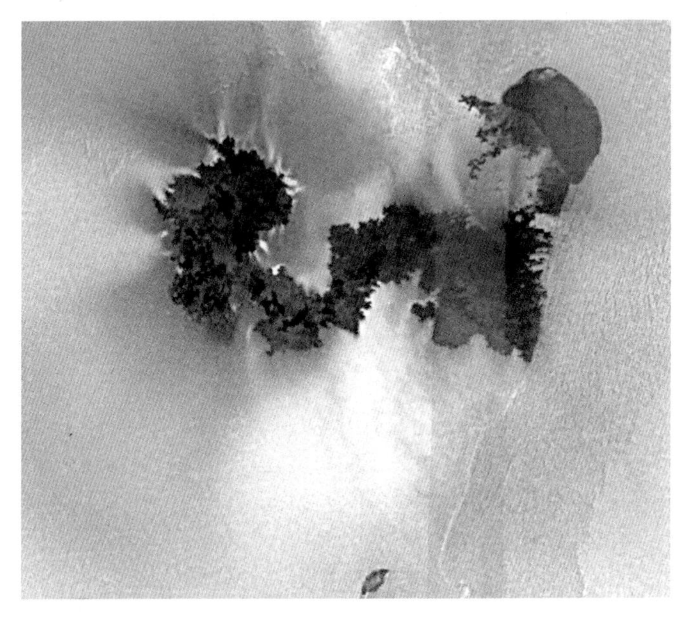

(43 mi) to the west. Researchers found this puzzling until they studied Galileo data and realized that the volcanic source of Prometheus had not actually moved, but the lava flow had. As lava moved across the landscape, it crusted over, probably flowing inside lava tubes. About 70 km (43 mi) downstream from the vent, the lava breaks out again, interacting with sulfur dioxide frost on the surface. This interaction is thought by some to give birth to the visible plumes that erupt today. Or the plume may be caused by the hot lava eroding the surface and interacting with frozen sulfur dioxide underground.

Plumes on Io are incredibly energetic, in part because of Io's low gravity but mostly because of the moon's lack of significant atmosphere. Volcanologist Susan Kieffer estimates that if Yellowstone's Old Faithful geyser was transported to the low gravity and near vacuum of Io, its terrestrial 50-meter-high (160-ft) gush would blossom to an altitude of 35 km (22 mi).

Io's third type of volcanic eruption is the most dramatic.

(*top*) The Galileo spacecraft captured this image of Io's Amirani volcano. The caldera and lava flow are slightly to the left of the center in this picture. At more than 300 km (190 mi) long, this is the largest known active lava flow in the solar system. (*bottom*) The jellyfish-shaped caldera of Io's Prometheus volcano is seen in the upper right of this high-resolution image made by the Galileo spacecraft. The lava flow that extends from the caldera across sulfur dioxide plains is about 90 km (56 mi) long.

(*top*) This high-resolution image of Prometheus's frozen lava plains and vent was taken by the Galileo spacecraft. The entire area is underneath the plume erupting from Prometheus. The lava may be vaporizing bright sulfur dioxide ice.
(*bottom*) Geysers usually erupt cyclically, with short bursts of activity. Geysir is the Icelandic geyser from which all others get their name.

This colorized image of Io's Tvashtar volcano region was taken by the Galileo spacecraft. The lava flow (*left*) inside the caldera is more than 60 km (37 mi) long.

Pillanian eruptions (named after the violent Pillan hotspot on Io) emit the highest known plumes in the solar system. One of the most powerful was found by accident. Eight months before the discovery, Galileo coasted over the Tvashtar Catena region. At that time, a curtain of glowing lava 22 km long (14 mi) fountained from a fracture in the floor of a great caldera. A nearby plume exploded gases more than 100 km (62 mi) into the void. The remarkable activity of Ionian volcanism became a concern for Galileo flight engineers a few months later. An encounter in August 2001 was to bring the spacecraft over the same region, into the path of the volcanic debris. Would the explosive jet of sulfurous particles damage it? Flight engineers decided to go ahead with the somewhat risky trajectory. When Galileo flew over the region, there was no detectable plume activity at the original site. Surprisingly, another volcano—a new one—blasted out of the ground about 600 km (370 mi) away. Galileo flew into a slurry of sulfur dioxide snow. This close encounter of the volcanic kind provided scientists with their first direct contact with Ionian material. As Galileo swooped within 194 km (121 mi) of the surface, it coasted through the new volcano's searing breath and survived, apparently unscathed. The 500-km-tall (310-mi) plume, the most powerful yet seen, was lofted by a volcano called Thor.

The Pillan volcano is the source of Io's mysterious high-temperature lavas. The temperatures of Io's ubiquitous eruptions give scientists insight into their nature and clues to the composition of the lava. Early Voyager data were

The volcano Pillan, Io's "black eye," rests on a ring of red material thrown out by the volcano Pele.

consistent with low-temperature sulfur-based volcanism (or cooled silicates), but two instruments aboard Galileo, the NIMS (Near Infrared Mapping Spectrometer) and SSI (Solid State Imaging system) were better equipped to measure temperatures. The temperatures measured at the Pillan hotspot by Galileo's robot eyes were surprisingly high. Earlier flybys had detected many high-temperature sources resulting from silicate eruptions, but no one had expected temperatures higher than terrestrial eruptions. Early in earth's history, high magnesium lava resulted in superheated eruptions. These ultrabasic eruptions discharged scorching lavas in the range of 1800K (1527°C [2780°F]). But Galileo data suggest temperatures in the 1997 Pillan eruption were above this mark. Does Io have ultrabasic magmas? The question remains open.

The largest observed eruption on Io to date was witnessed from earth through the Keck II telescope, high atop Hawaii's Mauna Kea volcano. A titanic explosion from the volcano Surt (named for the Icelandic volcano god) spread across the northern hemisphere of Io on February 22, 2001. Incandescent fountains of superheated molten rock burst forth, covering an area estimated to be 1,900 sq km (730 sq mi), an area larger than the entire city of London. With sophisticated telescopes using adaptive optics—bendable mirrors that correct for atmospheric blurring—ground-based observations will give us more and better data as time goes on.

One of the unsolved mysteries of Io is its extraordinary coloration. Most of its brilliant surface hues can be attrib-

uted to sulfur dioxide, which goes through dramatic color changes as it cools. Molten material is most often black, changing as it cools into red, orange, and yellow. Sulfur dioxide frost also powders the landscape in blue and white. But some areas, such as the crater of the volcano Culann, have been dubbed "golf courses." The Ionian golf courses are not as carefully tended as the fairways of Pebble Beach or Muirfield, but they are decidedly green. One of the most dramatic examples of the Ionian palette is a sprawling crater complex called Tupan (named after the thunder god of Brazil). Tupan is 75 km (47 mi) across, sporting a multicolored, fractured surface encircled by cliffs 1,000 m (3,000 ft) high. The widespread fissures give the appearance of lines drawn by a giant's black marker. Tupan has strips of the golf course terrain. The remarkable green hues may be caused by sulfur or by a volcanic mineral called olivine. Whatever its cause, Tupan's bright face is a microcosm of the beautiful, varied, and completely alien environment of Io.

In this artist's view of the interior of the Tupan caldera, sulfur melts down from the walls, leaving red and green deposits on the floor. Europa (*left*) and Ganymede are seen in the sky.

The volcanic activity on Io is inexorably connected to

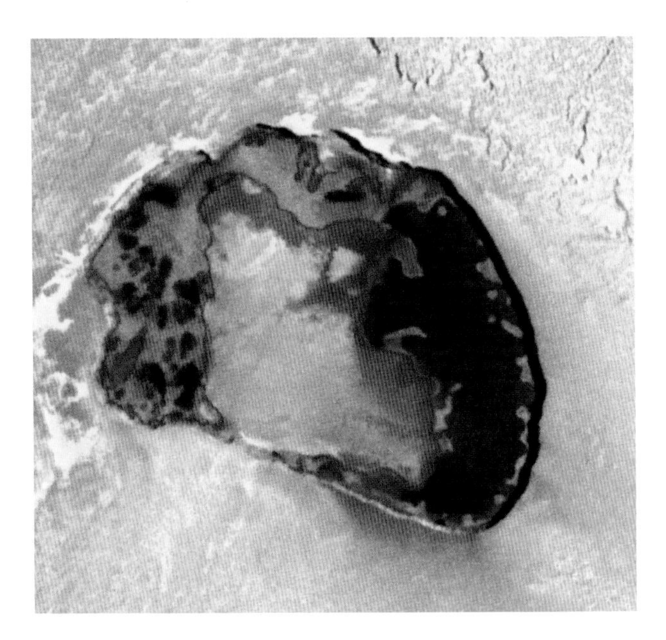

The Galileo spacecraft captured this image of the Tupan caldera. The caldera, about 75 km (47 mi) across, has an active floor (dark parts) on its eastern (right) side, while lava pokes through the floor in places on the western side. The caldera has an "island" in the middle that is covered with sulfur dioxide deposits.

Io's torus. The torus, a cloud of plasma, originates on Io and encircles Jupiter. Every second, some 90,000 metric tons (100,000 tons) of material erupt from volcanoes on Io. A large portion of that escapes into space, drifting around Jupiter like a gigantic doughnut. As atoms of sodium and sulfur dioxide expand away from the moon, they slam into Jupiter's electrical fields, becoming ionized, or electrically charged. Once this happens, a powerful sheet of electrical current known as the flux tube sets up a sort of electrical short circuit between Io and Jupiter. An astonishing 2 trillion watts of electricity course through the doughnut-shaped cloud, equivalent to the output of 350 nuclear power plants. The torus is just one of many intriguing features born of Io's violent and bizarre eruptions. Dante would have loved it.

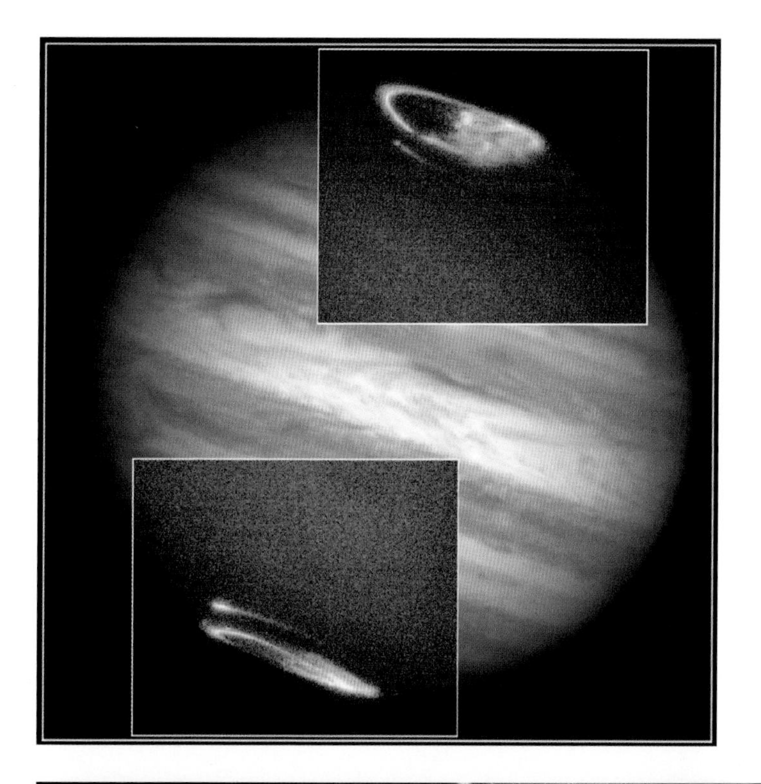

(*top*) Jupiter's auroras can be seen through the lens of the Hubble Space Telescope that orbits earth. The Jovian auroras are fed by the electrical fields passing between the planet and Io.
(*bottom*) If the human eye could see the titanic energy fields pulsing through the Jupiter system, we might observe a scene similar to this one. A tube of energy links Io to Jupiter, feeding the giant planet's auroras. Io travels in a superenergized torus of sodium as it circles Jupiter.

6 HINTS OF CRYOVOLCANISM

The volcanoes of the inner solar system and Io are forged in the furnaces of molten rock. But magma is not the only recipe for volcanic eruptions. Alien brews are simmering out there. Frigid gases escape from the moons Triton and Enceladus. Other strange concoctions are birthing eruptions of a very different kind, powered by "magma" of exotic chemistry: superchilled water mixed with ammonia, methanol, and other strange blends. These alien eruptions are called cryovolcanism. To power these watery eruptions, investigators think there must be substantial reservoirs of water beneath the surfaces of icy moons. Cryovolcanism is essentially the eruption of liquid or vapor phases of water, with or without solid fragments of materials, which would be frozen solid at the normal surface temperature.

The icy satellites of Jupiter, Saturn, Uranus, and Neptune were first investigated up close by the Voyager spacecraft. Water is the most important constituent of these moons with the notable exception of Io (see chap. 5). While many of these satellites' surfaces are peppered with impact craters, there are some apparently young, smooth areas. These areas are thought to have been resurfaced by cryovolcanism, though it is possible that other processes, such as ice tectonism, diapirism (the convection of solid ice), and intrusive volcanism (in which the "magma" doesn't break through to the surface) also played a part.

Cryovolcanism is not a process we see on earth, though features such as the "icefoots" on the shore of Lake Superior are similar to cryovolcanism. These icefoots are formed in winter, when waves generated over the open lake water

(*opposite*) In this artistic imagining, a fresh ice-filled crater on Europa reflects the face of Jupiter. Io passes to the right, trailing a cloud of sodium.

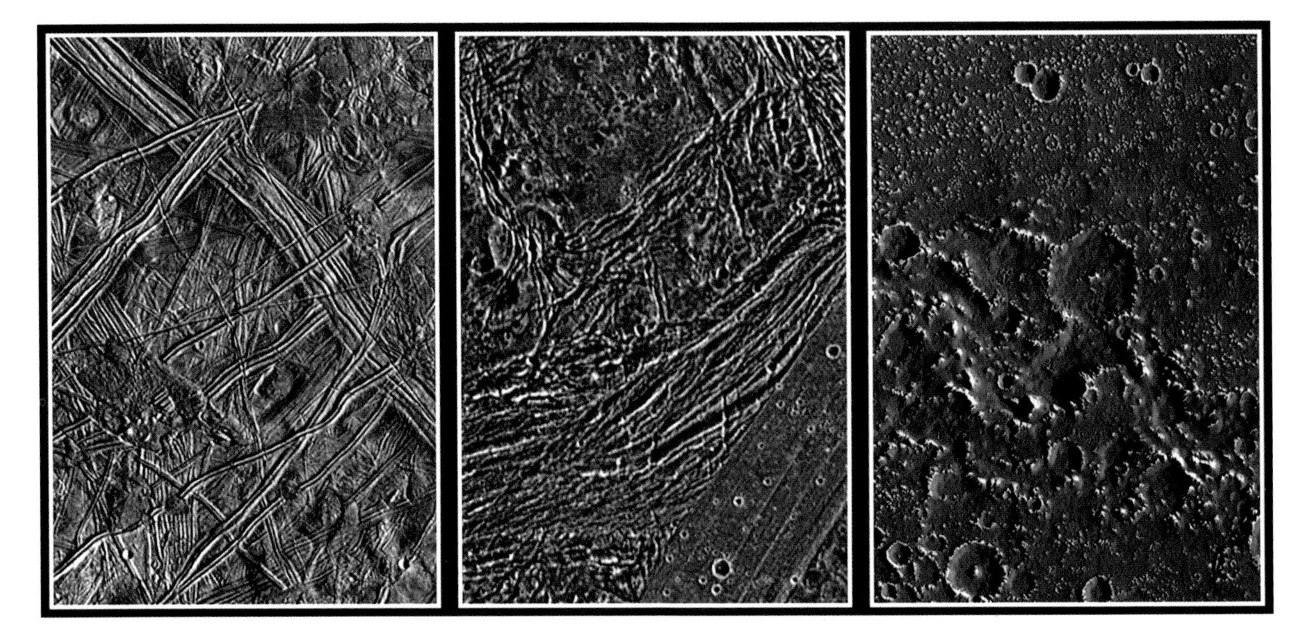

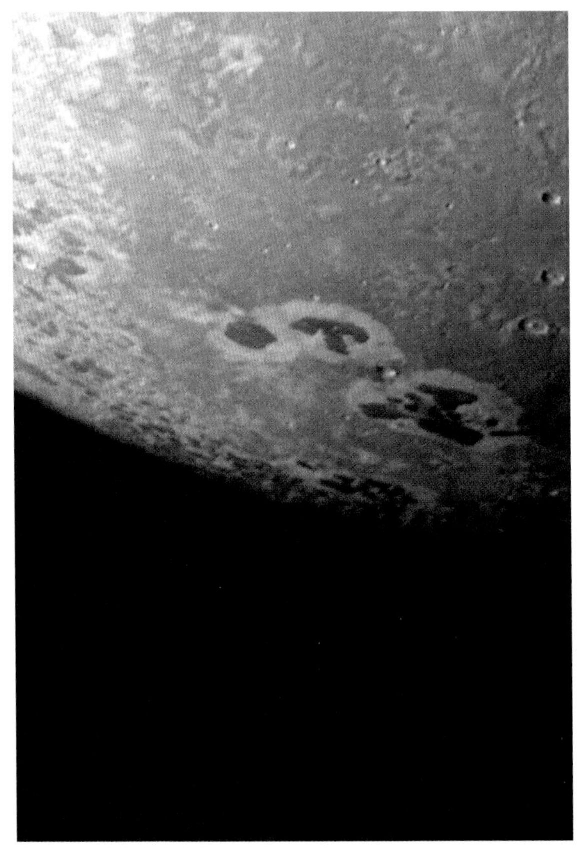

(*top*) The surfaces of Jupiter's moons Europa (*left*) and Ganymede (*center*) exhibit evidence of cryovolcanism, unlike their geologically older sibling Callisto (*right*). (*bottom*) The face of Neptune's moon Triton has been covered in places by cryolavas. This Voyager 2 image shows several irregular dark areas called guttae. These giant lobate features, which have a dark core surrounded by brighter materials, are thought to be cryovolcanic flows. (Image shows an area about 1,000 km [621 mi] across.)

pump liquid water through cracks in the ice at the lake's margins. The resulting ice mounds (icefoots) may be considered comparable to an expression of cryovolcanism around the "vents."

In order to have cryovolcanism, a moon or planet must have liquid water in its interior, and this liquid water, or mixture of water and other materials, must be able to erupt at the surface. A number of factors are important here, including the moon's size, the amount of radiogenic heating, and the history of the moon's orbit (whether tidal heating occurs).

Another important factor in cryovolcanism is the composition of the "magma," which is often referred to as "cryomagma." Water alone presents a problem, as liquid water is denser than ice. In theory, it is hard to erupt water through a solid ice crust (remember that ice cubes float). But cryomagmas are probably not made up only of water. Other materials, such as ammonia, could lower the melting point and the density of cryomagmas and make them easier to erupt. The composition of the cryomagmas will depend on the distance of the body from the sun and what materials condensed from the solar nebula at that distance. Rocky materials and metals tended to drift toward the inner solar system in its formative epoch, whereas lighter materials accumulated in the outer system, away from the sun's heat and wind. Researchers think that methane and ammonia condensed from the solar nebula at the distances of Jupiter and Saturn, so it is likely that these could be present in cryomagmas. In the outer reaches of the solar system, carbon monoxide, carbon dioxide, and nitrogen may play an even greater role.

As with the magmas of molten rock on earthlike worlds, the thickness (viscosity) of cryomagmas depends on its mixture of water and other materials. Liquid water would just flood a surface, filling in topographic lows. But a mixture of water and ammonia will have viscosity similar to that of silicate lavas, which means similar landforms would grace these distant landscapes: flows with tall margins, shields, and domes. We still have much to learn about how cryomagmas erupt, but given the discoveries by spacecraft like Voyager, Galileo and, more recently, Cassini, our expectations are high.

EUROPA

For most of the era of modern space exploration, the earth has been thought of as the "water planet." Perched between broiling Venus and freeze-dried Mars, the earth was seen as an oasis of liquid water, a damp outcast in a family of dry worlds. True, many moons in the outer solar system

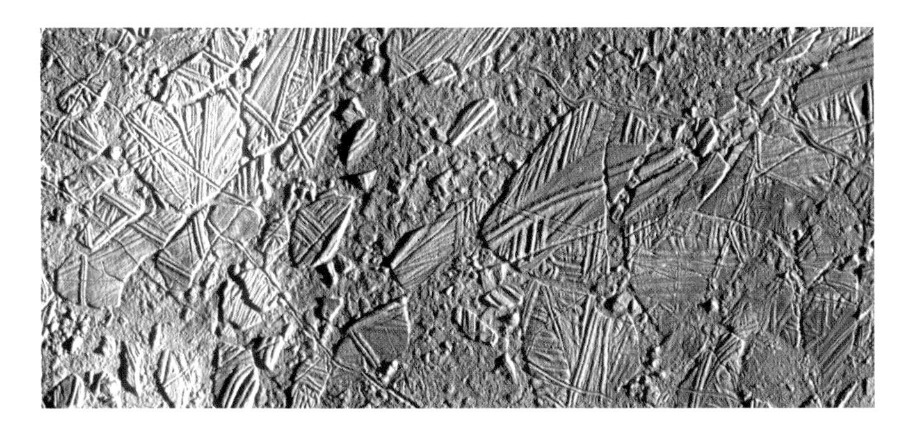

Regions like Conamara Chaos reveal how Europa's crust has fractured into floating rafts that then froze into position. This closeup view shows details as small as 9 m (30 ft) across. The scallops in the cliffs are about the size of the cliff face at South Dakota's Mount Rushmore.

appeared to be balls of ice, and the comets were known to be made up of frozen water and gases. But the earth had liquid water on its surface, something that even its close sibling, Mars, seemed to lack. Recent revelations about the Martian environment have tempered this view, but the contrast between transient flowing water on Mars and the massive water inventories on earth remains striking. Earth is, indeed, a water world.

Then came our first views of Europa from Voyager, followed by the Galileo spacecraft from much closer range. Jupiter's fourth-largest moon sparkles with a brilliant white surface made of shimmering water ice, blemished by few impact craters. Its pristine surface indicates a very young age, only about 50 million years by some estimates. An elegant calligraphy etches the surface in linear and arcurate stripes, telltale signs of deep fractures. Researchers constructed several models to describe conditions beneath Europa's bizarre, grooved facade, ranging from soft ice to an ocean 100 km (62 mi) deep. Whatever its true internal form, Europa is far more a "water world" than the earth. If surface eruptions occur, they must be understood in light of the hypothesized subsurface ocean.

High-resolution images taken by the Galileo orbiter are suggestive of an ocean on Europa. Glistening lines snake across the frozen landscape, bracketed by long ridges rising hundreds of meters into the black sky. The ridged surface has fractured into vast sections of ice called rafts. These rafts appear to have shifted and rotated before freezing solid again. Many of these rafts can be fit back together like a

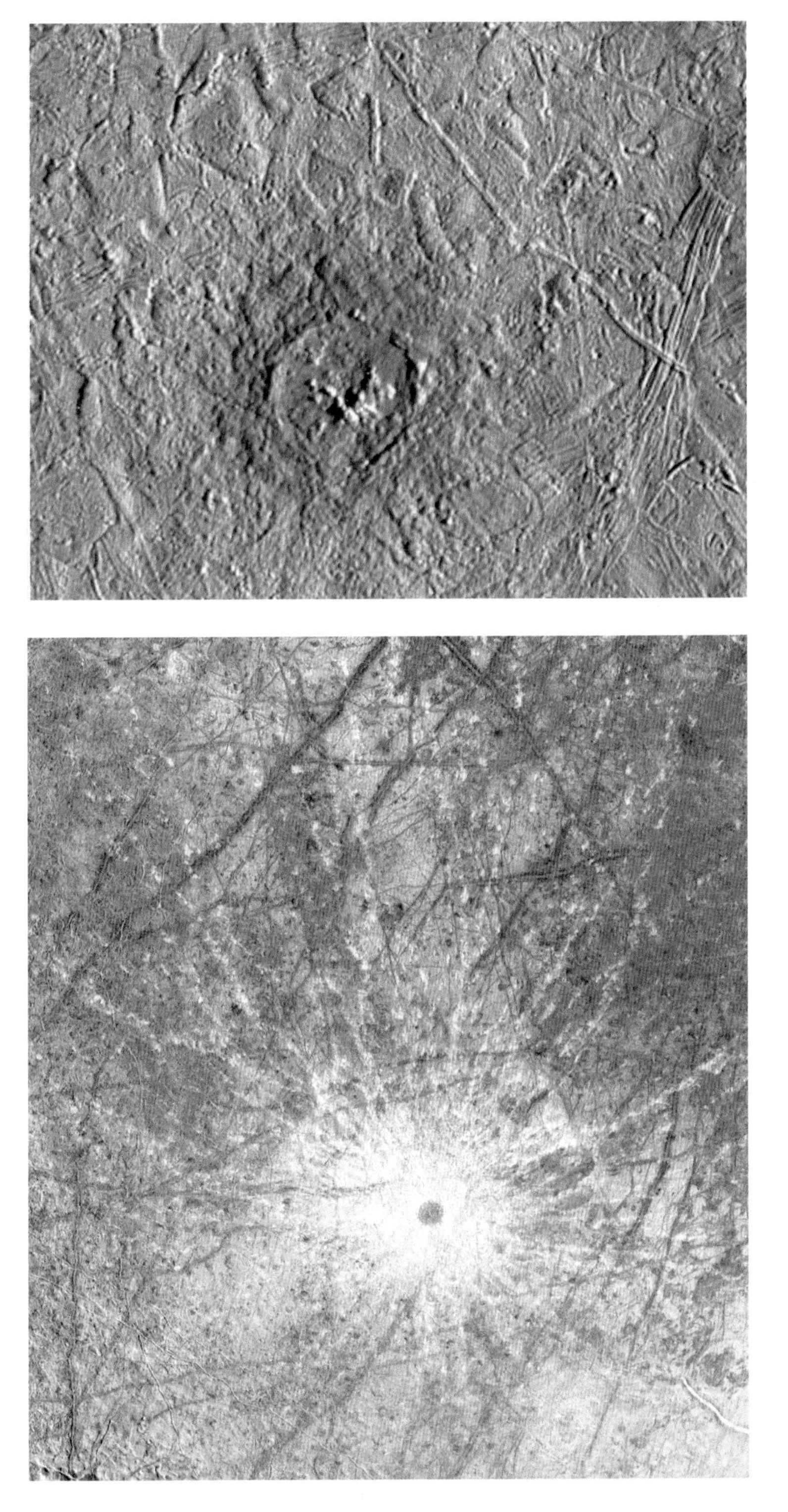

Pwyll is one of the few major impact craters identified on Europa. The meteor that gave birth to the crater apparently crashed through a thin layer of ice, allowing water to fill in after the impact. Rays of fresh ice cast out by the impact can be seen in the lower image. The crater is about 26 km (16 mi) across.

cosmic jigsaw puzzle, clearly indicating that a once continuous surface has been split up and moved around.

Aside from visual clues, Europa generates a magnetic field consistent with liquid saltwater. In the first week of 2000, the Galileo spacecraft flew within 346 km (215 mi) of Europa's surface. As the craft passed through radiation fields from nearby Jupiter, its magnetometer charted a change in magnetic fields coming from Europa. These directional changes were quite similar to those that would be generated electrically by conducting liquid within Europa's upper ice region. Unlike the electrical currents pouring from earth, Europa's field is induced—it is created in response to Jupiter's mighty field lines. This induced field is generated by, and continually changing in response to, the rapidly rotating magnetic field of Jupiter. Europa circles the giant planet in just over three and a half earth days, while Jupiter spins on its axis once every ten hours. Its magnetic field is tilted by nearly ten degrees from the spin axis, and Jupiter's pace swings its field around with it, sweeping through Europa. The laws of physics tell us that under these conditions, any conducting material swept by this field will create (or induce) a magnetic field to offset the rapidly changing external magnetic field. This induced field is exactly what scientists observed as the Galileo spacecraft flew past Europa. The presence of the induced magnetic field led researchers to conclude that a near-surface conducting layer, such as an ocean with dissolved salts, was the culprit. Currently, that's the best explanation for the observations.

The earth's magnetic north pole is at the top of the planet. Europa's magnetic north pole moves along its equator, in sync with the energy flowing from its parent planet. From Europa's perspective (because of its own movement), Jupiter appears to rotate once every eleven hours. The induced magnetic field from within the moon cycles to compensate for the changing Jovian magnetic field, which sweeps by twice each "day." A compass held by a Europan explorer would completely reverse direction every five and a half hours.

Europa's dynamic landscape and magnetic field present a conundrum: researchers estimate that a body the size of Europa—subjected to the frigid Jovian environment—should freeze solid within a few million years. While some heating could come from radioactive elements within the silicate core, this heat would not be enough to preserve a liquid layer. But another force is at work here, the same force that powers the volcanoes of Io: tidal heating. Europa makes half a circuit around Jupiter for each time that Io circles, and twice for each of Ganymede's orbits. In terms of gravitational pull, Europa is caught between Io's "rock"

This artist's rendering depicts Europa's induced magnetic dipole axis, which lies in the equatorial plane of Jupiter. A polar flux tube connects Europa to Jupiter, much as another tube links Io to the giant planet. The flux tubes are related to the "footprints" of the moons that can be seen in Jupiter's upper atmosphere.

and Ganymede's "hard place." Tidal heating, generated as a result of the moons' orbits, heats up Europa's interior, though to a much lesser extent than Io's. Conventional volcanism may exist on Europa's "seafloor," at the junction of the silicate mantle and the ocean. Europa may have a volcano-rich seafloor nestled beneath an oceanic crust.

But what of surface activity? Scientists suggest that the hidden Europan ocean may have given rise to ice volcanoes, geyserlike eruptions, or both. Dozens of sites on the moon hint at past eruptive events involving water that has rapidly frozen in Europa's near-vacuum environment. One of the clues leading to this conclusion lies within the "triple bands," highwaylike parallel stripes that contribute to Europa's cracked eggshell appearance. In Voyager images, structures within many Europan stripes, called linea, can be seen as bright lines running down the center of a dark, well-defined band. The bands are less than 15 km (9 mi) across, but run over Europa's face for thousands of kilometers. The stripes are directional, with bands in the northern

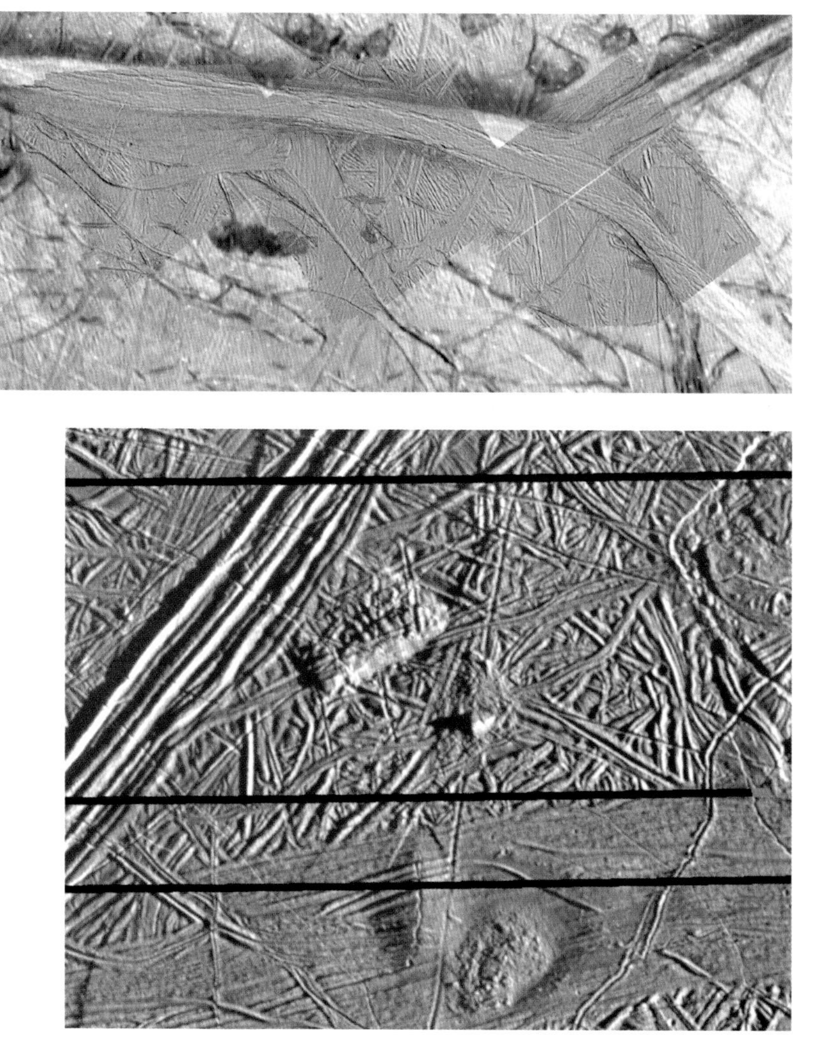

(*top*) Fissures like Rhadamanthys and Agenor Linea (seen here) provide circumstantial evidence of cryovolcanic eruptions on Europa. Agenor Linea splits into two sections at the right. This image covers an area about 135 by 60 km (84 by 37 mi). (*bottom*) This Galileo spacecraft view of Europa takes in a triple band at left, as well as mounds and depressions, all of which may be related to cryovolcanism or a subsurface ocean. The black stripes are areas where no data were returned. This image depicts an area about 32 by 40 km (20 by 25 mi).

hemisphere tending in a northwest direction, and southern bands tending toward the southwest. This directional tendency suggests a relationship with Europa's orbital stresses. Galileo's imaging system—superior to those of the earlier Voyagers—revealed the borders of the bands to be diffuse and irregular in many areas. Even the central bright median displayed patchy sites with haloes of bright material spilling across the dark outer band. The triple bands look like fissures erupting materials of varying albedos, or brightnesses, onto the moon's glistening surface. One such band is Rhadamanthys Linea, which lies across the surface of Europa like a beaded necklace.

A leading theory for triple band formation proposes a tidally induced fault breaking through the ice to the ocean below. A "cryolava" of briny water oozes up to seal the vent, while geyserlike eruptions emerge from weaker locations. This style of cryovolcanism is referred to as "stress-controlled cryovolcanic eruption." The water eruptions coming out of linear fissures may be analogous to rift-magma

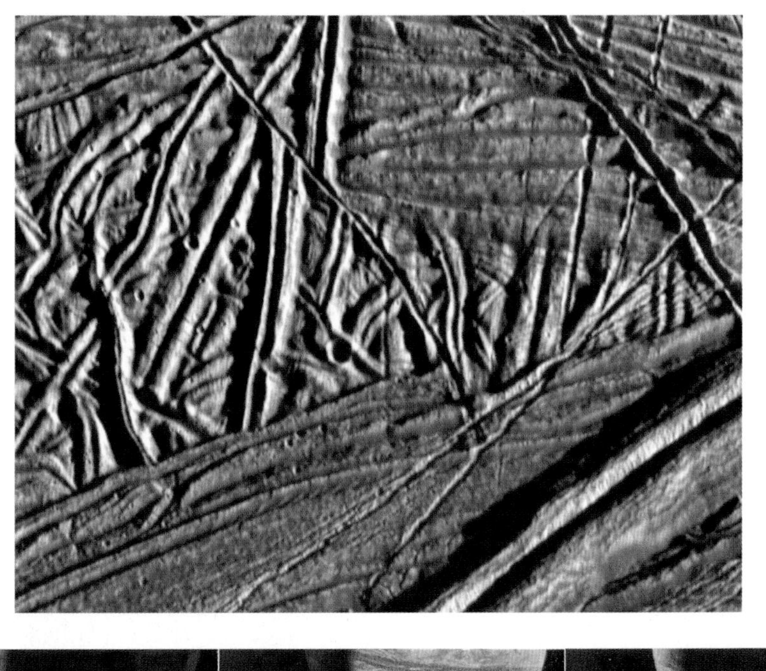

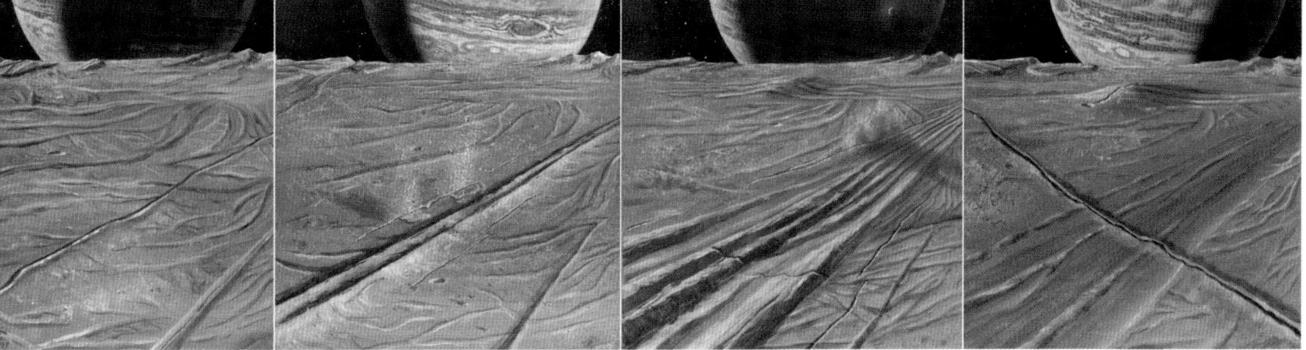

eruptions on Hawaii or to the fire fountains of Io's Tvashtar Catena. As the region around the fracture builds vertically, the weight of the growing ridge pulls on the surrounding ice, causing parallel fractures. These cracks, in turn, develop into more parallel ridges, duplicating the process as the band expands in girth.

Another theory proposes that the ridges mark boundaries of colliding ice plates. These compression ridges sink under their own weight. As the surrounding ice is pulled downward, the sunken troughs along the ridge fill with dark material. While the details of the triple band morphology are not yet understood, it appears that the ridges are similar to the mid-Atlantic ridge, where seafloor spreading builds new territory beneath the Atlantic ocean. Shared characteristics include fields of small hummocks scattered along parallel faults and troughs that form a central axis. Terrain on either side of the medial troughs appears to be spreading symmetrically away from the center.

Still another theory holds that linea were emplaced by

(*top*) Scientists are not sure how the complex ridges and fractures on Europa were formed. The area depicted is approximately 15 km (9 mi) by 12 km (7 mi). (*bottom*) An artist imagines one possible sequence of triple band formation, as seen from about 35 km (22 mi) above the surface of Europa.

solid ice rather than liquid breaking through the surface. Planetologists who ascribe to this theory cite several reasons including the fact that many of the bands rise hundreds of meters above the surrounding plains. This suggests ductile ice rather than liquid, which would not result in a raised structure. The genesis of the hummocks also makes more sense in this scenario as well. Furthermore, there seems to be no flooding of material into adjacent ridges and valleys, which would have occurred had the linea been filled with liquid.

How thick is Europa's crust? Some studies suggest a thick ice shell essentially solid down to many tens of kilometers. Other models posit an ice crust of something like 10 to 15 km (6–9 mi). The structure and thickness of Europan crust is a hotly debated issue, and a complex one. Some researchers assert that impact craters and jumbled "chaotic zones" indicate a thin, 2-km (1-mi) crust at the equator. The crust appears to thicken to the north and south. Much of the evidence indicates that Europa has a global ocean, a vast subsurface sea running from pole to pole under the ice. Scientists were able to piece together a picture of Europa's interior by the way Europa's gravity affected Galileo's path. As the spacecraft sped up, its signal shifted, just as the siren on a passing fire truck shifts. The varying structure of Europa's crust caused subtle changes in this Doppler shift, enabling scientists to chart the moon's internal structure. Doppler data from the closest flybys fit a rocky interior capped by an outer shell of water 100 to 200 km (62 to 124 mi) deep. A 6-km (4-mi) crust is a fragile film over such a massive ocean.

While water ice is the dominant material on the surface, at least one other dominant material seems to be mixed in with the ice, and some of the data suggest that this material may come up from the ocean below. The amorphous, often radial discoloration may be the fallout from geyserlike activity. Such stained sites are called "painted terrain." Many such features exist on Europa. Most are associated with fractures or faults. The brown stains appear to be endogenic (generated from the inside), but whether they are localized events or global in nature remains to be determined. Some have well-defined flowlike, or pooled, edges. Others are diffuse, fading at the boundary. Often, dark material blankets the surrounding terrain, more like particulate matter than a flood of liquid. What is this dark material? Instruments aboard the Galileo spacecraft gave researchers enough data to identify several candidate substances. Investigators studying Europa's infrared (heat) data suspect various salts, with magnesium sulfate (like Epsom salts) as the best spectral match. Other investigators argue that Europa's spectra is

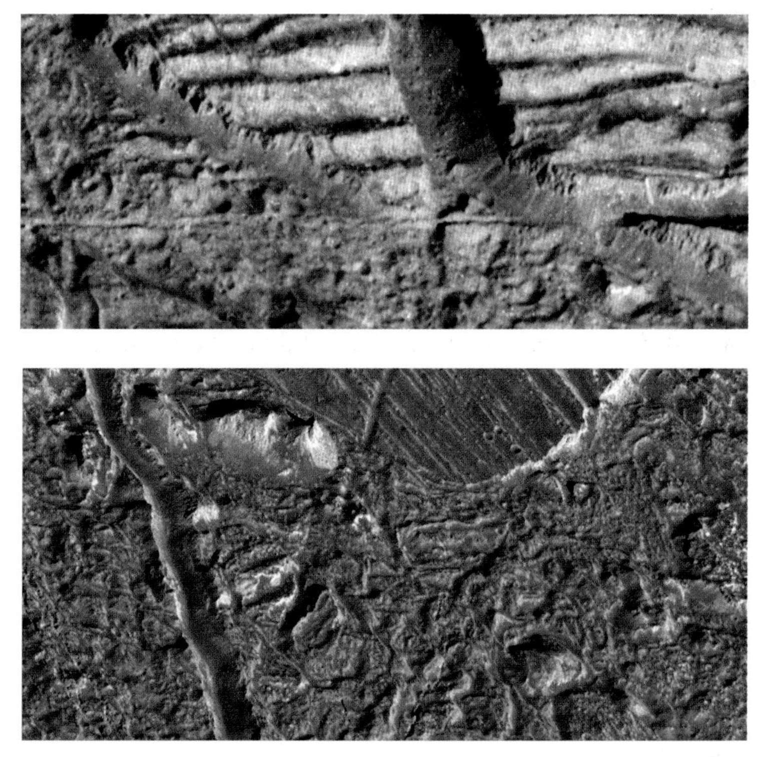

Images of Europa at the highest resolution show ice flows and rugged terrain. *(top)* Corrugated plateaus end in icy cliffs over 100 m (330 ft) high. Image resolution is 9 m (30 ft) per picture element. *(bottom)* Terrain in the Conamara Chaos region. Image resolution is 9 m (30 ft) per picture element. *(right)* Highest resolution image of Europa, 6 m (19 ft) per picture element.

best matched by hydrated sulfuric acid. While salts are not brown, sulfur is. The inner Jovian environment is bathed in sulfur, thanks to Io's eruptions. Other scientists have suggested iron compounds that presumably issue from the rocky core.

Cryovolcanism may also be exhibited in ponded and flooded areas. Cryolavas appear to have erupted or seeped onto the surface leaving frozen pools. These smooth areas embay low-lying terrain, oozing into adjacent valleys and troughs before freezing solid. The erupted material is darker than the surrounding landscape and may be the expression of briny subsurface lakes.

Given Europa's orbital stresses, many planetary geologists think it likely that Europa's ocean floor is pockmarked by volcanic vents. If current models are correct, those vents are submerged in as much as 100 km (62 mi) of water. The great distance might preclude any direct disruption of the surface, so Europa may not exhibit any visible clues about seafloor volcanism. Still, mysterious features hint at forces

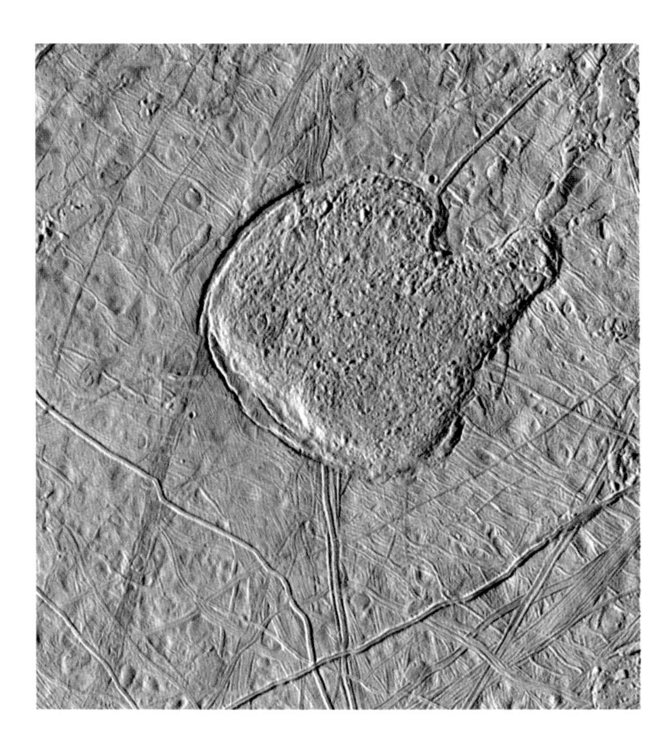

This mitten-shaped depression on Europa has fractured into jumbled, chaotic terrain, perhaps from a heat source below. The image depicts an area about 175 by 180 km (109 by 112 mi).

beneath the crust. The most striking of these are the chaotic regions. Here, ridges are seen to slip across each other in lateral faults, while remnants of ridged ground have broken and rotated in a slush of debris. Chaotic terrains, which often coexist with relatively smooth, craterless landscapes, bear a striking similarity to sea ice in arctic regions. In the terrestrial case, solid ice has fractured, drifted into new positions, and been frozen in place again.

Some researchers suggest that plumes of hot water, generated by seafloor volcanic sources, could thin the ice, eventually triggering a series of melt-through events. The ice would fracture, freeing rafts of surface ice to bob and rotate in the quickly solidifying lake. If the concept is correct, the rafts should rotate in the same direction as the hypothesized plume (clockwise in the north). This seems to be the case, although data are limited.

It is also possible that chaotic regions have been generated more indirectly. One concern is the distance between Europa's ocean floor and the base of its surface crust. If models are correct, plume material would take weeks or even months to migrate from the seafloor to the surface. Additionally, chaotic regions would need to be heated for extended periods of time to explain the crustal movements observed. Some researchers have solved the distant plume problem by proposing warm columns of ice. In this scenario, a heated plume warms the ice over a long period. The heat moves through the ice much as the Day-Glo material in a lava lamp. Rather than melting completely through the ice, the process would be gradual and relentless, softening

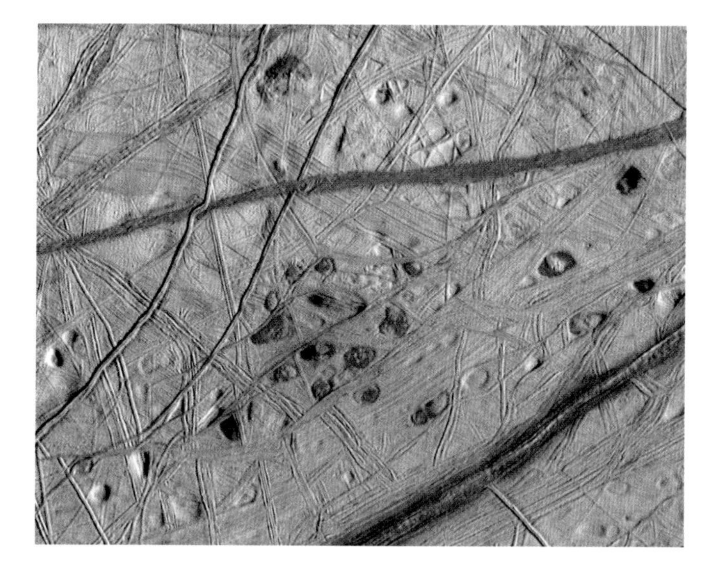

The Galileo spacecraft captured this area of lenticulae, or dark spots, on Europa. Lenticulae is Latin for "freckles." Each dark spot is about 10 km (6 mi) across.

the ice enough to free the rafts for extended periods of time. Impurities in the ice might also help the process along by lowering the melting point. Small amounts of salt or sulfuric acid—both of which have been tentatively identified by Galileo—could help the icy mixture move upward, even through tens of kilometers of ice.

One widely accepted scenario posits that while the ice is heated from beneath—possibly from seafloor plumes—no melt-through occurs. Instead, a rising solid mass (called a diapir) makes its way through the crust, eventually reaching the surface. These diapirs could also interact with pockets of trapped briny water. Chaotic regions need not be generated quickly, but could be the result of a long and gradual process.

The diapir model would help to explain the lenticulae, another feature that may reflect seafloor volcanism. Lenticulae are areas that come in varied forms, including miniature chaos areas, pits, depressions, and domes. Lenticulae, Latin for "freckles," refers to the fact that most of these features are dark. Many of these spots tend to be depressed beneath the surrounding plain. The ruddy ice is briny, bolstering the idea that dark materials are seeping—or erupting—onto the surface.

The upwarped domes that comprise some lenticulae span a diameter of roughly 10 km (6 mi). Supporters of the volcanic plume theory point out that although long columns of heated liquid would expand and dissipate by the time they reached the ice, they may contain gases that dissolve at the base of the ice crust. These gases could be responsible for the domed structures. Many of the domes appear in the vicinity of the chaotic regions. It has also been suggested that they result from vortices and eddies spinning off a main plume. But the ice diapir model is more consistent with the similar size of all lenticulae. If solid state convection is taking place within Europa's crust, warm diapirs would generate consistent scales of lenticular features.

The frozen wasteland of Europa still has secrets hiding under its surface. Are there submarine volcanoes lurking beneath 100 km (62 mi) of saltwater? Are such geothermal hotspots analogous to anything on earth?

Not until 1977 was seafloor volcanism discovered on earth—on the ocean floor along the Galápagos rift zone near the Galápagos archipelago. Explorers had noted hot undersea plumes, but their nature and source remained a mystery until the deep-sea submersible Alvin revealed dramatic chimneys of sulfur compounds rising from the ocean floor. Gradually, researchers realized that the number of volcanoes on the ocean floor must far exceed the five hundred to six hundred active ones on the surface. Undersea hydrothermal vents tend to occur along the midocean ridges where new crust is being created in the conveyor-belt style of seafloor spreading unique to earth. While undersea volcanism is not limited to plate boundaries, the midocean ridge areas may well be the most volcanically active sites on earth. Vents tend to cluster in groups, much as they do in Yellowstone. The earth's mantle comes to within a few hundred meters of the surface in these eternally dark regions. Seawater percolates through the earth's crust, eventually making contact with the 1200°C (2200°F) magma. The water becomes remarkably hot (up to 540°C [1000°F]), as the high pressure prevents it from boiling. The heated fluid makes its way up through fissures in the rock, leaching minerals along the way. When it finally streams into the ocean, it is laced with a complex mineral soup. Mineral-rich water erupts from these sources, building delicate structures of spires and chimneys, some of which may tower dozens of meters above the seafloor. The streams of water are often laced with materials that lend names to their forms. Some vents are known as black smokers, while others are called white smokers.

Perhaps the most remarkable phenomena associated with these ocean volcanic sites are the colonies of life huddling around them. Biologists had long thought that the ocean floors were sterile. Far from the sunlight that feeds earth's complex web of life, the cold depths were seen as a sort of desolate, high-pressure desert bathed in eternal darkness. But the sulfur and other minerals carried by hydrothermal vent eruptions provide nourishment for an entire

(*top*) Visitors to Hawaii stand at the seashore where lava flows into the water. Volcanoes enrich ocean water with minerals from deep within the earth.
(*bottom*) Seafloor volcanism was discovered on the ocean floor in 1977. Seawater percolates through the earth's crust, eventually making contact with superhot magma. The heated fluid makes its way up through fissures in the rock, leaching minerals along the way. Mineral-rich water erupts from these sources, building delicate structures of spires and chimneys. This black smoker was discovered on the Juan de Fuca ridge near the U.S. West Coast.

biome. Sulfur is taken up by bacteria that are completely independent of any food sources related to solar energy. Some of the bacteria associated with black smokers, thought to be similar to the most ancient life on earth, are known as archeobacteria. These microbes provide the foundation for an alien menagerie of Pompeii worms, blind crabs, giant tubeworms, one-eyed shrimp, and other strange beasts living in these frigid pressure-filled environments.

The deep-sea colonies at hydrothermal vent sites have invigorated the astrobiology community. Scientists who study the possibility of life on other worlds have not missed the possible analogy between the terrestrial ocean floor and conditions that may exist beneath the oceans of Europa. Europa may provide a salty brine to seafloor volcanic sources triggered by tidal friction. As life exists independent of the sun in earth's deep sea gardens, it may also have taken hold on this distant, watery world.

GANYMEDE

Orbiting farther away from Jupiter than Europa, Ganymede carries the distinction of being the largest moon in the solar system. With a radius of 2,634 km (1,637 mi), it is larger than the planet Mercury. It is also the only moon known to generate an internal magnetic field, implying a hot convecting core containing iron. Ganymede may also have an induced magnetic field, suggesting a deep internal ocean of liquid briny water.

Ganymede's geology is complex, displaying a marked contrast in its terrains: dark terrain, heavily cratered, forms about one-third of the surface; the other two-thirds consist of swaths of bright, grooved terrain that may have been formed by both cryovolcanic eruptions and tectonic forces. In the aftermath of the landmark Voyager Jupiter encounters at the end of the 1970s, the consensus was that Ganymede was frozen in its evolution. The moon seemed to have stabilized somewhere between the paths taken by Europa and Callisto. The dark, ancient terrains resemble those of Callisto, while the bright regions are somewhat like Europa. But with new data from the Galileo mission, we realize that Ganymede is more complex, its history not so simple.

Ganymede's dark terrain is heavily cratered and thought to be ancient, older than 4 billion years. Its composition shows a greater fraction of rocky material than the bright terrain. However, geological investigations using Galileo high resolution images indicate that the dark material is a relatively thin layer that overlies brighter, icy materials. The dark terrain has been modified by numerous geologic processes, including tectonism, but evidence for cryovolcanism is not conclusive.

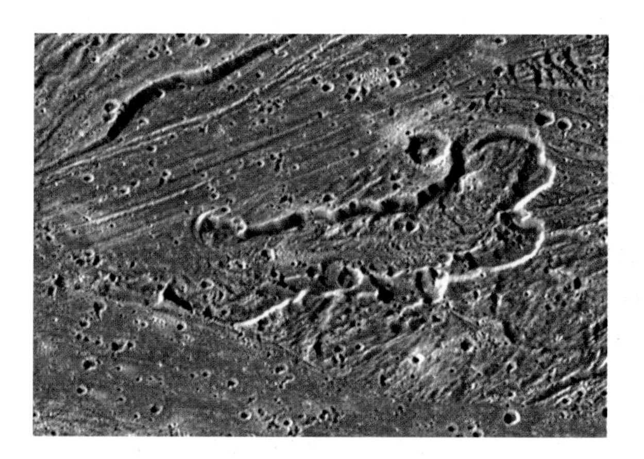

A lobate flow covers the floor of this caldera-like feature in Ganymede's Sippar Sulcus region. The depression is 55 km (34 mi) wide.

The composition of Ganymede's bright terrain is almost pure water ice, suggesting cryovolcanism, but it also shows ridges and valleys, indicating that tectonic forces were at play. The bright terrain, sometimes referred to as grooved terrain, is very complex to interpret. We don't know its origin with any certainty or why it covers only two-thirds of Ganymede. The bright terrain may be a result of resurfacing by cryovolcanism, tectonism, or both. However, even with the high spatial resolution images returned by Galileo, how volcanism and tectonism might have contributed to the formation of the bright terrain is still not clear. No flows have been identified in the images, but there are some arcurate depressions that could be the sources of flows, perhaps similar to volcanic craters or calderas on other planets. These caldera-like features may be sources of some of the bright terrain materials. This question remains unanswered for now, as do many others. We don't know why the bright terrain is located in swaths that crisscross the moon, or why the whole moon was not affected by the resurfacing. Ganymede remains a mysterious and bizarre world, with much for future researchers to find out.

MIRANDA

While Ganymede's geological forces may not yet have fallen silent, Uranus's diminutive moon Miranda exhibits evidence of long-dead cryovolcanism. Miranda, which is only 470 km (290 mi) in diameter, was remarkably active in its past. Many mysteries remain, as less than half the satellite has been seen in detail. That hemisphere is sculpted in three

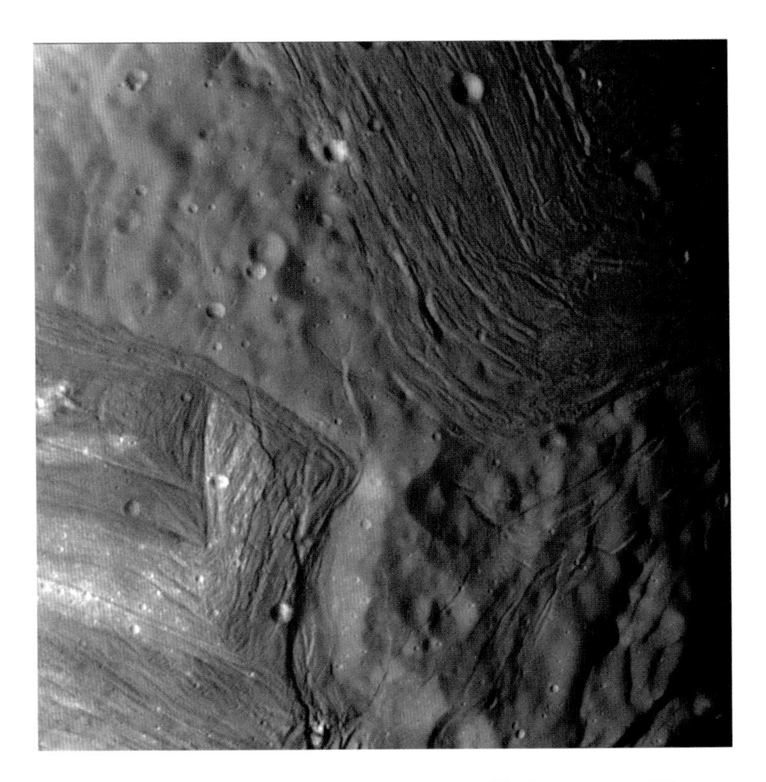

Miranda's bizarre face has been sculpted by internal forces, as evidenced by this part of Inverness Corona with its pale chevron (*lower left*). Material appears to have flowed to the surface from an internal heat source. The area depicted covers about 220 km (137 mi).

areas by wrinkled, polygonal zones called coronas. These chaotic regions are up to 300 km (190 mi) across, consisting of concentric ridges and troughs. While some ridges appear to be faults, others resemble volcanic flows that may have been a mix of ammonia and water.

Scientists think that the coronas formed over plumes of rising material from Miranda's core. As material spread out and impinged the surface, they triggered volcanism. There are features indicative of flood deposits oozing from fractures at several sites. Tidal heating may be responsible for the internal heat, as Miranda interacts gravitationally with Uranus and nearby sibling moons. The Uranian moon Ariel also exhibits some evidence of resurfacing and cryovolcanism.

TITAN

When the two Voyager spacecraft flew by the Saturn system in 1980 and 1981, they returned spectacular images of many moons, but those of Titan revealed nothing of the surface. Titan appeared as an orange ball, its surface covered by a thick atmosphere.

Saturn's largest moon is one of the most exotic and mysterious bodies of the solar system. Discovered in 1655 by the Dutch astronomer Christian Huygens, Titan guarded its secrets well until the Cassini spacecraft arrived in 2004. Titan has the second densest atmosphere of any of the solid bodies we know, second only to Venus. This atmosphere shrouded the moon, completely veiling its surface during the Voyager 1 close flyby in 1980. Much to the disappoint-

ment of planetary scientists at the time, all that Voyager's camera showed was a fuzzy orange globe.

Titan has long been thought to be a likely place for volcanic activity. At 5,150 km (3,200 mi) in diameter, Titan is larger than Mercury and second only to Ganymede among moons in the solar system. Titan's substantial mass and density suggest that plenty of gravitational and radiogenic energy is available for melting its interior. In addition, its eccentric (noncircular) orbit around Saturn provides some tidal friction, though much less than Io's. Even before the arrival of the Cassini orbiter, researchers thought Titan might show cryovolcanic features on its surface. Various models of Titan's interior called for a substantial layer of water-ammonia liquid underneath an icy shell. If there was enough fracturing, the liquid could erupt through to the surface, causing cryovolcanism.

Another reason to think that Titan may have cryovolcanic activity is the atmosphere. Titan's thick atmosphere is about 95 percent nitrogen and a small percentage of

Lurking beneath the orange fog of Saturn's moon Titan, Ganesa Macula may be a massive cryovolcano with flows of water-ice on its flanks as depicted in this artist's imagining.

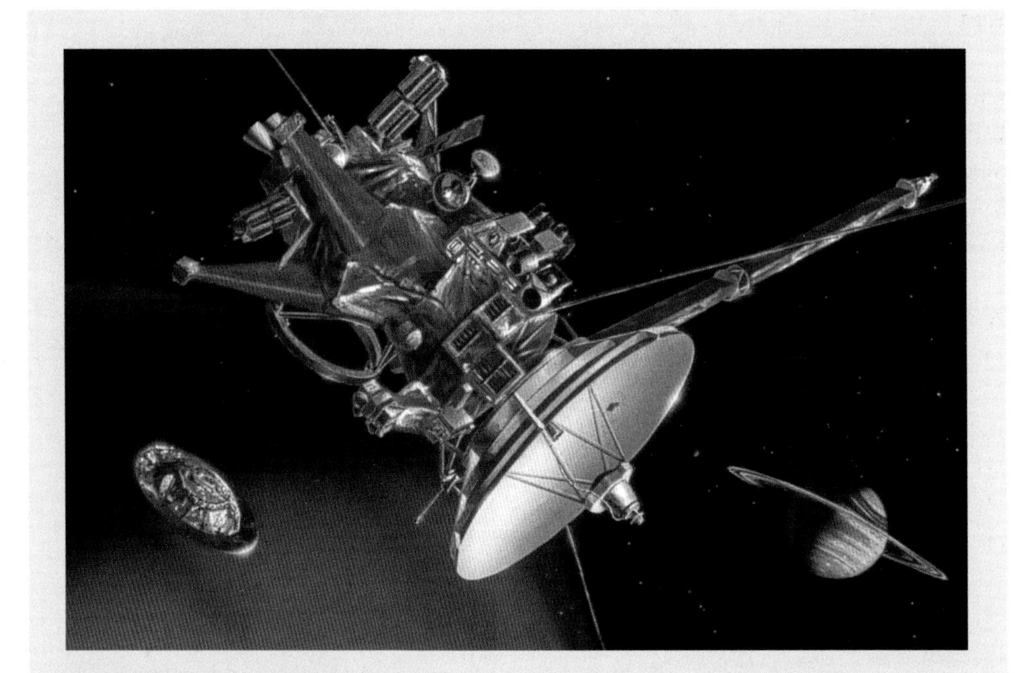

SATURN EXPLORERS

The planet Saturn has been visited by four spacecraft: Pioneer 11, Voyager 1 and 2, and Cassini-Huygens. Pioneer flew by Saturn in 1979, snapping images of the planet and its rings and moons. Voyager 1 and 2, equipped with more sophisticated instruments, encountered the golden giant in 1980 and 1981. The Voyagers discovered many moons and sent back detailed information about the Saturnian rings and weather. In 2004 the school-bus-size Cassini spacecraft arrived at Saturn. This was the first mission dedicated exclusively to the study of the ringed world. Cassini also carried the European Space Agency's (ESA) Huygens probe.

This artist's rendering depicts the Cassini spacecraft (*right*) and ESA's Huygens probe in space. Saturn is visible in the lower right; its moon Titan looms in the background. Bristling with twelve state-of-the-art scientific instruments, Cassini settled into orbit around Saturn on July 1, 2004. The Huygens probe, which landed on Titan on January 14, 2005, remained active for more than two hours and studied the atmosphere and surface of the moon. Cassini is scheduled to make at least forty-five flybys of Titan and dozens of closeup studies of Saturn's other moons. The Cassini-Huygens mission is the culmination of decades of Saturn exploration by spacecraft.

methane. Methane is photodissociated in Titan's atmosphere; that is, the sun breaks methane down so that the gas recombines with other atmospheric constituents, forming organics like ethane, propane, and acetylene. This means the methane is somehow being replenished. One thought was that large lakes, perhaps even an ocean, of methane or ethane could be resupplying the atmospheric methane. At Titan's temperatures (a very chilly 97K [–176°C (–285°F)] at the surface), methane behaves much like water on earth.

Titan has a few areas of liquid, but we now know it has no major surface ocean.

Another possibility for replenished methane is cryovolcanism, which may supply methane and other gases to the atmosphere. Some researchers think Titan's atmosphere started out as ammonia that erupted from the interior. The ammonia was later converted to nitrogen by sunlight, with leftover hydrogen escaping to space.

Scientists expected Titan's cryovolcanic features to be somewhat different from those on other satellites such as Triton and Europa because of Titan's thick atmosphere. The atmosphere has three effects. First, atmosphere makes it harder for gases to come out of the cryomagma. Second, it affects how far away from the vent any explosive products land (they land closer in thicker atmosphere). Finally, cryolavas cool faster in denser atmosphere. Researcher Ralph Lorenz predicted that cryovolcanic eruptions on Titan would more likely be effusive than explosive, not only because of the atmospheric pressure (which puts a damper on expanding gases) but also because there simply wouldn't be much gas in the cryomagmas. Some researchers expected Cassini to discover domes similar to the pancake domes on Venus. They also predicted that cryomagmas would likely be a mixture of water ice and ammonia, possibly with some methanol. Laboratory experiments suggested that these cryomagmas would be quite viscous, perhaps the consistency of wet concrete.

As the data flowed back to earth from distant Saturn, Cassini's results suggested that cryovolcanism has indeed been a significant geologic process on Titan and may be a major contributor to the atmospheric methane. On October 2004, Cassini executed its first targeted flyby of Titan, turning on its synthetic aperture radar (SAR) to peer through Titan's orange haze. Cassini's first radar pass obtained images at spatial resolution of about 350 m (1,150 ft), covering about 1 percent of Titan's surface. The most prominent feature revealed in these radar images was a circular feature that scientists interpreted as a cryovolcanic dome. Later named Ganesa (after the Hindu god of good fortune), the feature is about 180 km (110 mi) in diameter, larger than the domes on Venus. It could be a shield rather than a dome; since we do not yet have topographic information, we cannot know for sure. In general, a shield is made up of many thin, low viscosity flows, whereas a dome is formed by more viscous flows. If we establish the shape of Ganesa, we will also have clues to the composition of the cryomagma.

Other features that appear to be cryovolcanic in origin are also visible in SAR images. Several large flows spread

The most prominent feature revealed in Cassini's first radar pass over Titan was Ganesa Macula, a large circular structure (about 180 km [110 mi] across). Radar dark (smooth) at the top and radar bright (rough) on the edges, it is thought to be a dome or shield volcano.

across Titan's frigid landscape. They are most likely cryovolcanic, but some could be fluvial. Titan's surface shows plenty of branching channels, indicating that rivers of liquid methane have run there. Methane may fall to the surface as rain. Perhaps there are methane monsoons, periodic torrential downpours of superchilled liquid. Rain, if the accumulation is enough, can cause flows on the surface. Cryovolcanism can also cause flows, so the challenge is to identify which process caused a particular flow deposit.

Some of the flows seen in the radar images are much more likely to be from cryovolcanic activity than from rainfall, especially those that appear to come out of craters. The craters themselves are elongated rather than circular, indicating origin by collapse (volcanic) rather than by impact. The association of flows (in one direction only) with nonimpact craters is hard to explain by any process other than volcanism.

Subsequent Cassini flybys using radar showed features that may or may not be cryovolcanic. Since many flybys are still to come, we can expect to gain a better understanding of Triton's makeup. To date, no features have been identified that appear to have resulted from explosive eruptions (such as plume deposits or steep-sided volcanic mountains).

Other instruments on Cassini also returned data suggestive of cryovolcanism. Images from the camera, although hampered by Titan's orange haze, showed a varied surface with light and dark regions. The spacecraft was able to use infrared wavelengths to penetrate the methane haze, showing detail on the surface. One of the early high resolution

MONSOONS ON TITAN

In 1975 science fiction author Arthur C. Clarke coined the term "methane monsoons" in his book *Imperial Earth*. Decades before Cassini-Huygens uncovered circumstantial evidence of such alien rains, Clarke described methane storms on the surface of Saturn's moon Titan. The story, which takes place in 2276, is seen through the eyes of Duncan Mackenzie, a human who was born on Titan. Although some of the details in *Imperial Earth* are outdated, Clarke just may have gotten the methane monsoons right.

images showed a feature now called Tortola Facula that may be cryovolcanic. The snail-like feature, about 30 km (19 mi) across, could have been formed by a series of viscous flows piling up around and on top of each other.

We do know that Titan has a remarkably varied geology. Using crater counting to determine age, researchers have concluded that Titan's surface appears to be quite young; only three impact craters have been found so far. Radar images also show ridges that may be tectonic, large channel-like features probably formed by the flow of liquid, and dark, smooth regions that appear to contain liquids. The camera showed a lakelike feature close to the south pole, and the radar instrument revealed numerous lakes near the north pole that are probably filled with liquid methane or ethane.

The Huygens probe, which landed on Titan on January 14, 2005, sent back data indicating that cryovolcanism may have happened on Titan. Although the surface images did not depict any features that were obviously cryovolcanic,

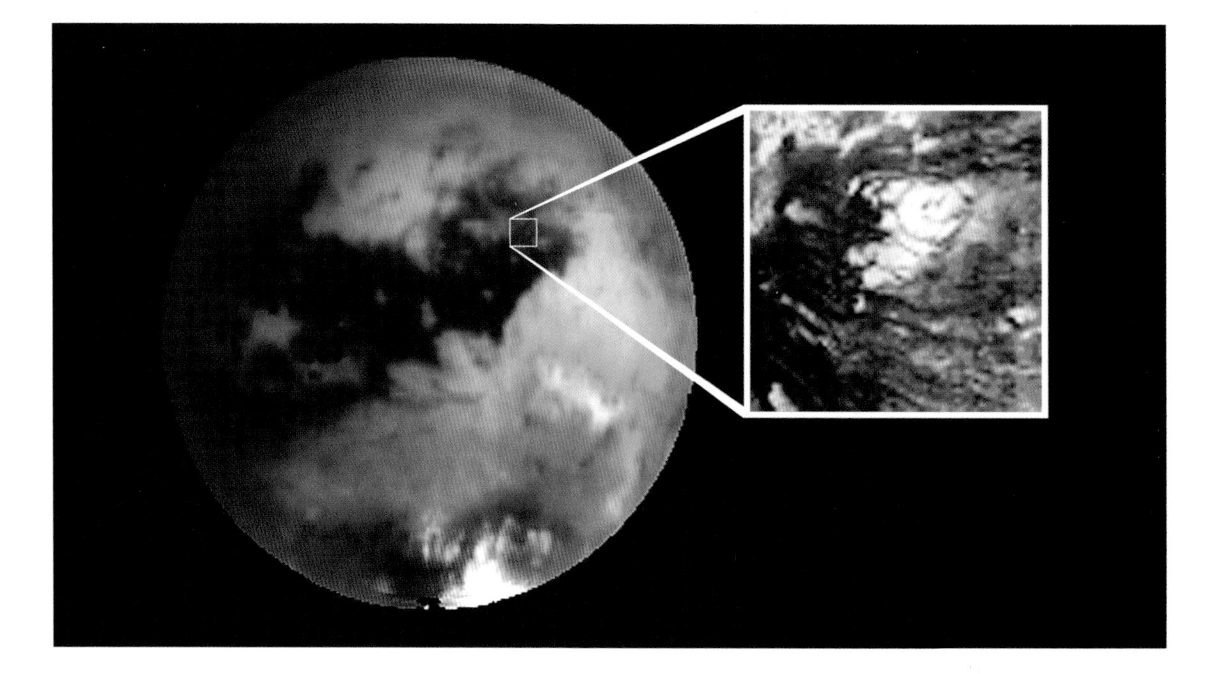

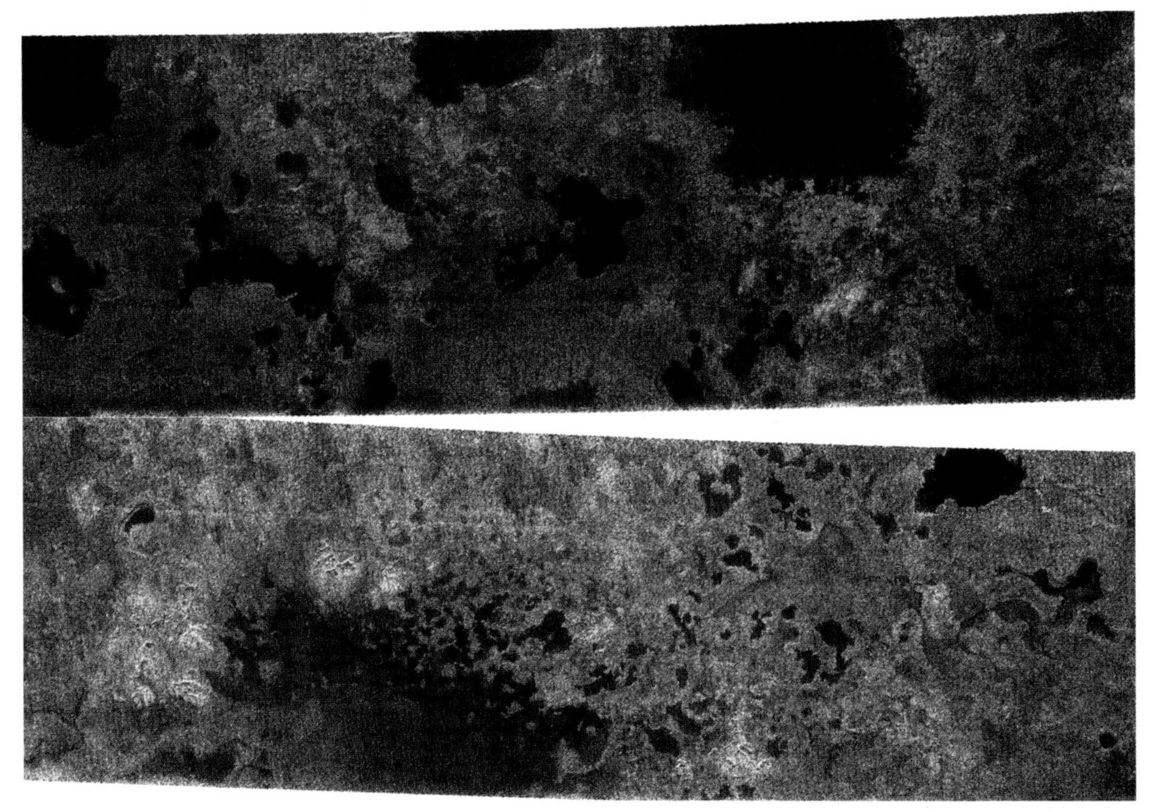

(*top*) This false-color image of Titan was made using three different wavelengths from Cassini's visible and infrared mapping spectrometer (VIMS). The inset shows a snail-like feature (named Tortola Facula) that may have been formed by cryovolcanic flows.
(*bottom*) Cassini radar images revealed dark patches that resemble terrestrial lakes sprinkled across the high latitudes surrounding Titan's north pole. Some of the lakes have channels leading in or out of them, with shapes that imply that they were carved by liquid. Radar studies suggest that several are filled with liquid today. The top image shows an area about 420 by 150 km (261 by 93 mi); lower image, 475 by 150 km (295 by 93 mi).

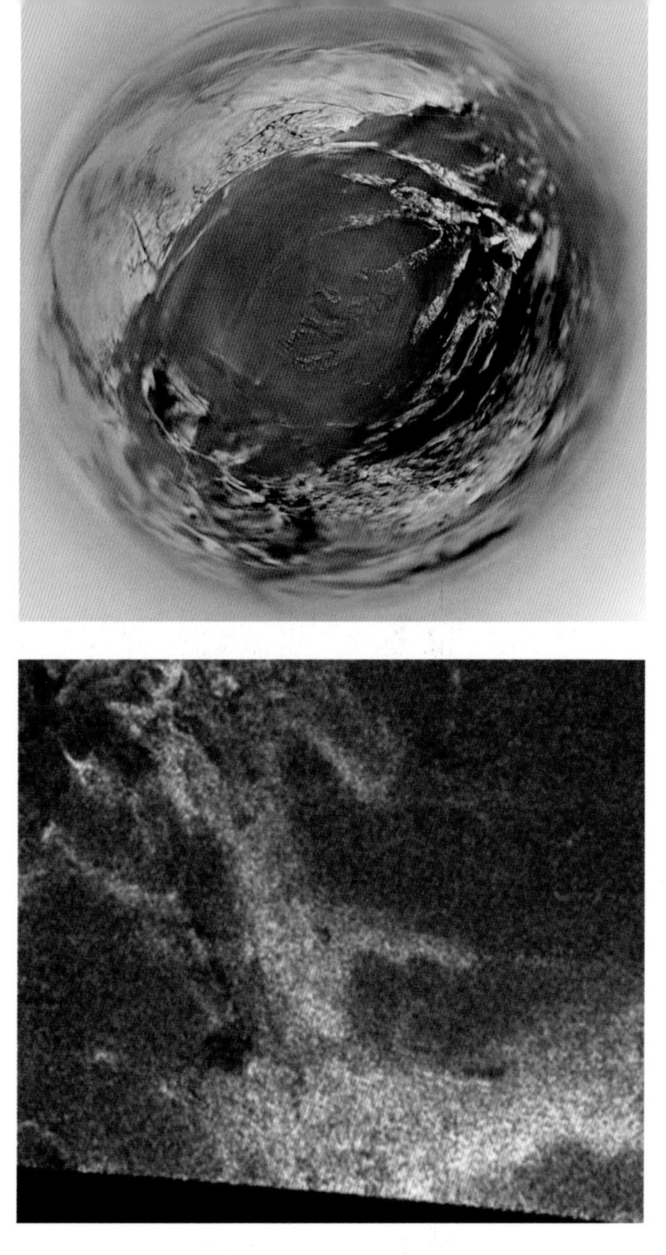

there was a surprising finding after the probe landed on the surface. Huygens found a specific type of argon (^{40}Ar) in Titan's atmosphere, indicating that Titan is bringing material to the surface from deep within itself. Cryovolcanism would be one means by which this material might be brought to the surface.

The big question at the moment is whether cryovolcanism still takes place and if Cassini's instruments are likely to detect any evidence of such activity. There is certainly plenty left for Cassini to find, but suggestions for cryovolcanic activity don't come from spacecraft data alone. Some ground-based observations show localized clouds outside the south polar region, leading to the suggestion that they may be related to volcanic eruptions from the icy interior. Some researchers think large cryovolcanic plumes might be releasing sufficient methane to account for Titan's atmospheric composition. As Cassini continues to observe Titan, more of this mysterious moon's secrets will be revealed.

(*top*) This fish-eye view of Titan's surface was taken by the Huygens probe from an altitude of roughly 5 km (3 mi). (*bottom*) Cassini radar captured this image of a cryovolcanic flow on Titan. The image covers an area of about 150 km (93 mi) square.

In this artist's rendering, geysers drape dark blankets of material across the frozen landscape of Neptune's moon Triton. Pink nitrogen ice melts directly into gas in Triton's near-vacuum, leaving behind strange eroded shapes.

7 PLUMES ON TRITON AND ENCELADUS

Once upon a time, volcanologists had no proof that earth was not the only place in our solar system where active volcanism occurred. Then Voyager 1 reached Io. Pictures and other data sent back to earthbound scientists provided incontrovertible proof of active volcanoes on that Galilean satellite. As space exploration continued, Voyager 2 showed us active geysers on distant Triton, and Cassini detected plumes on tiny Enceladus. Nothing is more exciting to a planetary volcanologist than the discovery of active volcanism on another world.

TRITON

Triton is the coldest and most distant moon so far visited by a spacecraft, and it might seem an unlikely place for volcanic activity. But evidence clearly indicates cryovolcanism on Neptune's largest moon.

Triton was Voyager 2's last stop before it went on its way to the outer edge of the solar system and beyond. Voyager 2 flew above Triton's south pole on August 25, 1989, taking stereo images that showed two dark, tall plumes, reaching approximately 8 km (5 mi) above the surface and leaving trails for about 150 km (93 mi). Other images of Triton's southern polar region revealed more than a hundred dark, streaky deposits, presumably a result of other plumes, implying that plume activity must be fairly common. We don't know how widespread this strange volcanic activity is on Triton, as the northern polar regions were in darkness during the Voyager flyby. At lower latitudes, images showed

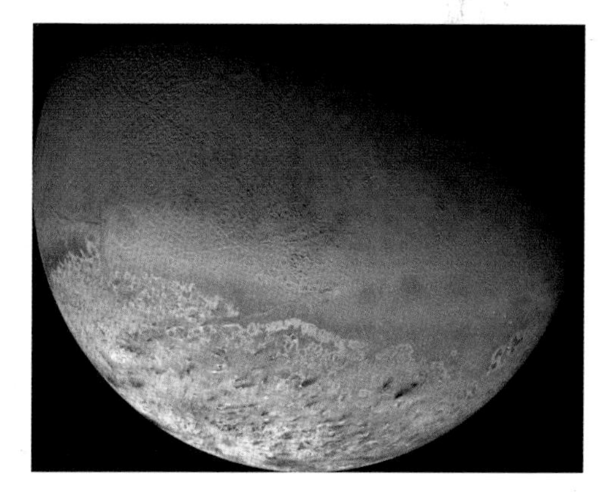

A global view of Triton taken during Voyager's 1989 approach to Neptune's largest moon.

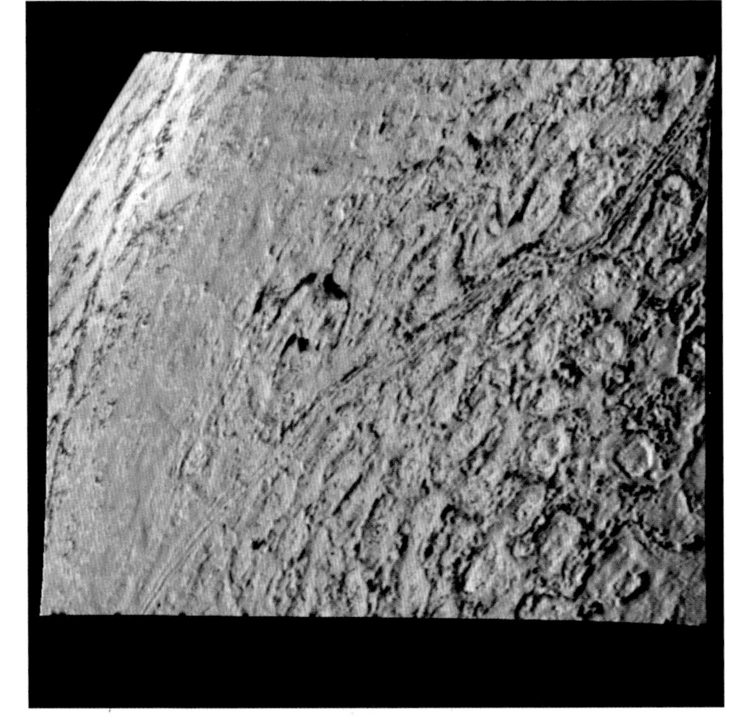

(*top*) Evidence of collapse and flow suggest that the smooth central area in this image of Triton may be cryovolcanic in origin. The small impact crater in the center is 15 km (9 mi) wide.
(*bottom*) This view of Triton's northern hemisphere is one of the most detailed images returned by Voyager 2. The circular depressions and raised ridges of the landscape may have been caused by local melting of the surface from internal heat sources. No other terrain like it has been found in the solar system.

a peculiar terrain with flowlike features, probably resulting from cryovolcanic flooding of older topography. This terrain has been dubbed "cantaloupe" because of its resemblance to that melon's skin.

Triton is a peculiar satellite. And cryovolcanism there is very different from cryovolcanism on the satellites of Jupiter and Saturn. Triton's orbit is inclined to Neptune's equator, and it is retrograde (opposite to Neptune's spin direction). Since nearly all moons circle their parent planets

Voyager sent back images of active plumes and the trails of gas they send above Triton's surface. These plumes may be similar to geysers. The arrows point at the horizontal trail of the plumes 8 km (5 mi) above the surface.

in the same direction as the planet's spin, any moon, such as Triton, that orbits in a different direction is suspected of being a captured object that formed somewhere else. Triton is very, very cold—only 38K (–235°C [–391°F]) at the surface. This is well below the freezing point of nitrogen, the material that makes up Triton's south polar cap. Both nitrogen and methane have been identified spectrally on Triton's surface, and carbon monoxide and carbon dioxide have been detected in minor amounts. Water ice is likely to be present in the crust, but it has not yet been detected. Triton has a very thin atmosphere made up largely of nitrogen. The atmosphere transports nitrogen ice from pole to pole every Triton year, keeping the surface temperature nearly the same everywhere.

We don't have any evidence that Triton's flows are still active, but the Voyager images showed active plumes. These plumes are thought to be similar to geysers, caused by solar sublimation of the polar cap, that is, melting directly from ice to vapor. Triton's southern hemisphere was just approaching summer solstice when Voyager flew by. This is a rare event as Triton's year is the equivalent of 165 earth years. Moreover, Triton's orbit is not round, but oblong. Because of this, it takes a wobbly path around Neptune, much as a top bobs as it spins. This orbital bobbing is called "precessing." Triton's orbit precesses every 688 earth years, causing the latitude of the warmest spot on the moon to change every Triton year. The region receiving the most sunlight on Triton wanders as far as 55 degrees from the equator. Voyager happened to fly by in one of those extreme

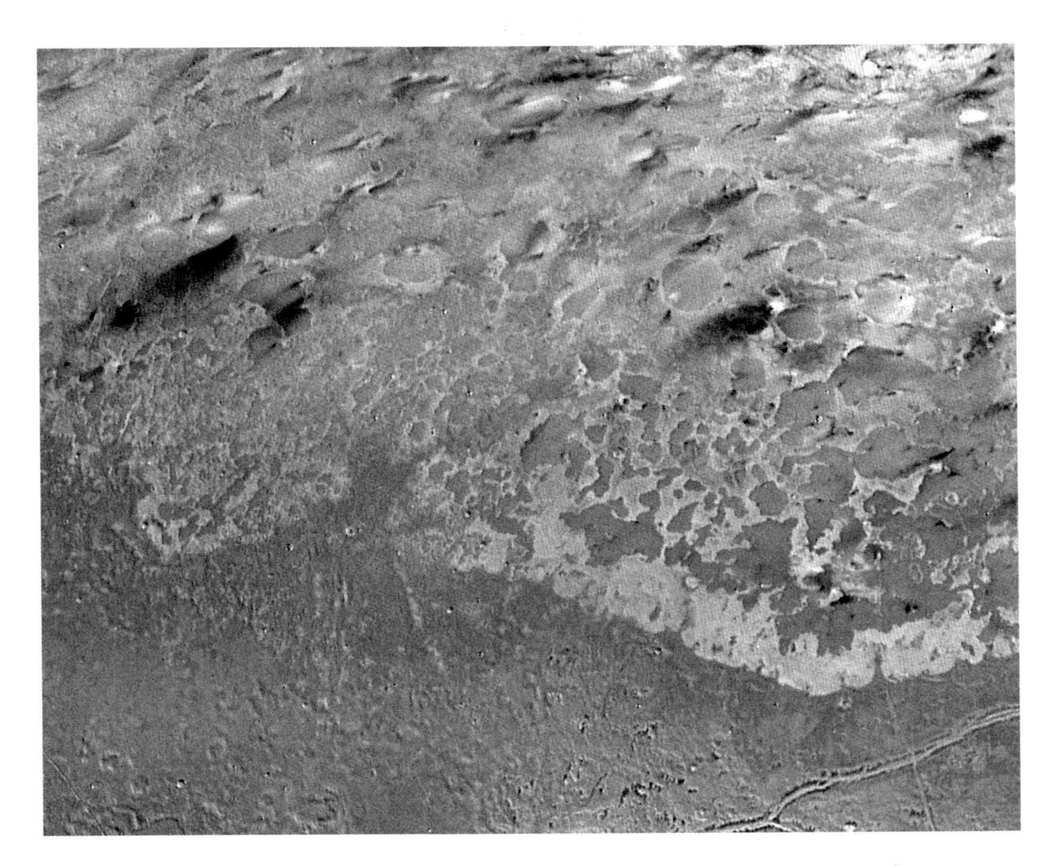

Dark streaks on Triton's pink nitrogen-ice surface were the first clues of active volcanism on the small moon.

years, when the sun was beating down on the rarely illuminated polar wastes. The active plumes were located close to the area where the sun was directly overhead.

One model postulates that the plumes are the result of sunlight beating down on and penetrating a transparent layer of low thermal conductivity nitrogen ice. This process has a terrestrial analog: the greenhouse effect. The greenhouse effect is not limited to particle-clogged atmospheres; it can operate within a solid, especially when that solid is nitrogen ice. Nitrogen ice is very clear, so sunlight can penetrate deep into the ice, creating a "solid-state greenhouse effect." Sunlight is absorbed and trapped by dark, carbon-rich impurities a few meters below the surface. This mild heating is enough to cause the interior of the nitrogen ice to become gas because an increase of about 10K (10°C [18°F]) above the surface temperature is enough to raise the vapor pressure of nitrogen by a factor of 100. The expanding gas then explodes into the near-vacuum of Triton's environment. If this model is correct, cryovolcanism on Triton is a side effect of sunlight, rather than an internally driven phenomenon.

There is much yet to be learned about this strange, frozen world, including its origins. Triton may have come from the Kuiper Belt, a band of comets and rubble that ranges from outside the orbit of Neptune to a distance of more than 6 billion km (4 billion mi) from the sun. While a new mission to Neptune is not yet scheduled, many scientists think the Neptune system is one of the highest priorities for future exploration.

ENCELADUS

The smaller bodies of the outer solar system exhibit volcanic activity when acted upon by larger bodies, especially those following eccentric orbits. The gravity fields of parent planets and sibling satellites cause an internal flexing that leads to tidal heating, a gravitational taffy pull that sometimes leads to volcanism. But at the close of the 1980s, few researchers would have guessed that another moon, riddled with volcanic activity, was lurking within the rings of the planet Saturn.

The golden giant Saturn has the most spectacular rings in the solar system. While Jupiter, Uranus, and Neptune also have rings, Saturn's are the biggest and brightest. From the inside to the outside edge, is approximately 323,478 km (201,000 mi), the equivalent of sixty-seven coast-to-coast trips across the United States. Scientists identify these beautiful rings by letters. Rings are lettered in the order of their discovery; the brightest central ring is B and the rings on either side of it are A (discovered at the same time as B) and C. Scientists knew for some time that Saturn's tenuous E ring was being continually replenished from an unknown source. They were also aware that the tiny moon Enceladus orbited within the ring, and many suspected that the ring's fine particles emanated from the moon itself.

Enceladus's pristine ice surface is, in places, a tortured jumble of twisted ridges and cracked plains nearly devoid of craters. These plains appear to have been resurfaced, with some areas having a geological age of less than 200 million years. Other parts of the surface are heavily cratered. All surfaces of Enceladus—ancient or young—are very bright, suggesting that the entire moon is dusted with fresh material.

A scant 504 km (313 mi) across, its diameter would cover the country of Spain. Because of its diminutive size, Enceladus's geologically young surface was mystifying. Other nearby Saturnian moons had cold, dead surfaces, despite the fact that many are larger. The orbit of Enceladus is not circular, similar to that of Io, so researchers assumed

The Cassini spacecraft took this edge-on image of Saturn's ghostly E ring. The view also captured two small moons, Calypso and Helene.

that tidal forces might be strong enough to trigger some kind of activity. The problem was that the moon next door, Mimas, also has an irregular orbit and wears a geologically quiescent, ancient face. Some planetary geologists suggested that the interior of Enceladus might be heated by a bobbing of the satellite, which may be caused by the tug of nearby moons.

When the Cassini spacecraft rocketed into Saturnian orbit in 2004, it found that the environs of Saturn were inundated with atomic oxygen. As Cassini coasted through Saturn's magnetosphere, it detected changes in the magnetic field lines. These changes revealed that ions from Enceladus were shifting the structure and shape of Saturn's magnetic fields. Instruments aboard the craft detected an atmospheric coma—a cloud of water, carbon dioxide, and methane—extending some 4,000 km (2,500 mi) from the surface of Enceladus.

Positive identification of cryovolcanic activity on Enceladus came from a trio of flybys from February to July 2005. The first encounter studied the frozen satellite from as close as a 1,000 km (620 mi). This nearly equatorial flyby returned detailed images of Enceladus's fractured surface. Folded, sinuous mountain chains, similar in scale to the Appalachian range in the United States, bordered darker plains in the south. These plains were laced with organic material. Ridges and grooves deformed the topography, hinting at crustal slippage and deformation. The magnetometer detected ions streaming from the moon's rarified atmosphere.

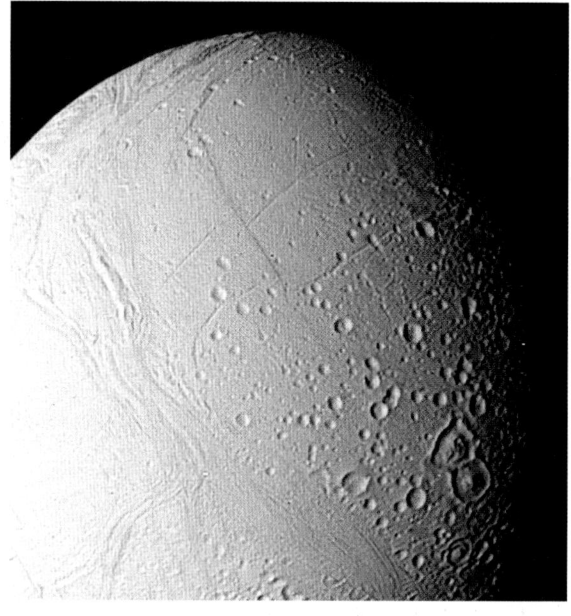

(*top*) Saturn's tiny moon Enceladus is shown in relation to Spain. The moon's diameter is only 504 km (313 mi). (*bottom*) One of the most detailed views of Enceladus was taken during the Voyager 2 flyby in 1981. Views like this, though limited, revealed mysterious flows on the ice moon.

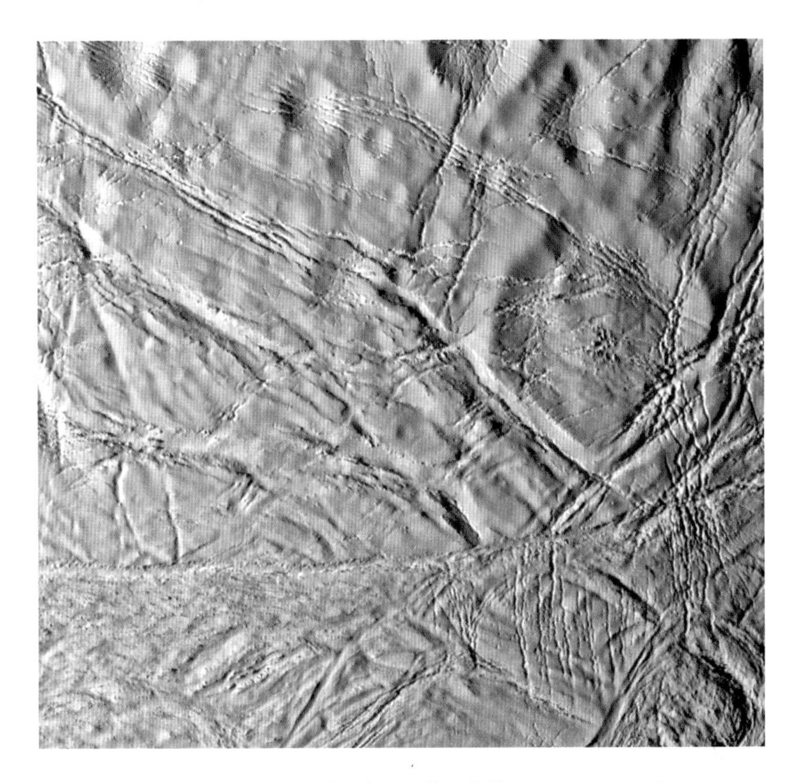

Some two decades after Voyager's encounter with Enceladus, the Cassini orbiter sent back more detailed images of flows and fractures on the small moon.

The second flyby took place the following month at half the distance. Magnetometer team members narrowed down the source of Enceladus's ion stream to somewhere in the southern hemisphere. Detailed images revealed what appeared to be surface flows. In some locations, these flows were diverted by low-lying hills. The ice movement was either similar to glaciers or perhaps a remnant of thick cryolavas. One image showed a possible cone with a caldera-like hollow at its summit. The case for some kind of active volcanism on Enceladus was building.

For the third flyby, Cassini's orbit was modified. The new course carried the craft within 168 km (104 mi) of the surface on July 14, 2005. Team members wanted more detailed data on the magnetic fields; they got more than they asked for. Cassini was able to fly directly through an extended plume of material. The spacecraft detected water vapor, carbon dioxide, methane, trace amounts of acetylene and propane, and possibly carbon monoxide and molecular nitrogen.

The spacecraft was also able to train its robot eyes on the star Bellatrix, in the constellation Orion. Cassini's path happened to bring Enceladus between itself and the star. Cassini carefully studied the fading light of the star as it passed behind Enceladus. When it did, Cassini detected the existence of a water vapor plume pouring from the south polar regions.

With the new data in hand, researchers estimate that Enceladus is losing 150 kg (331 lb) of water to space each second. While the material may not escape at a steady pace,

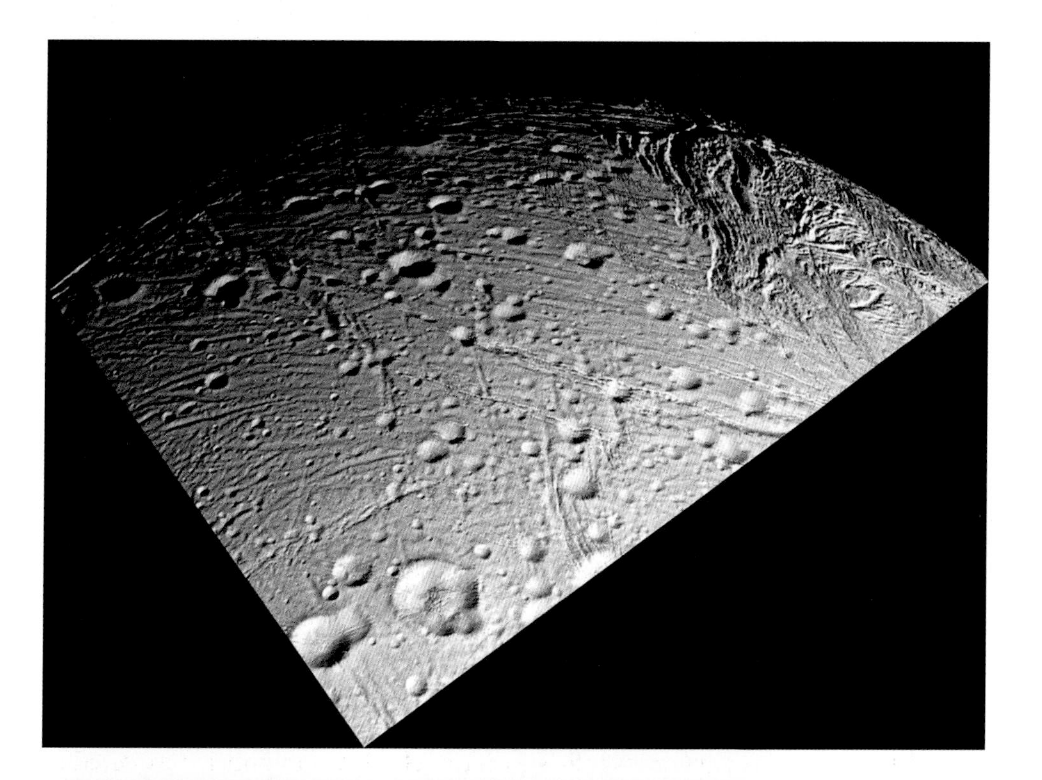

(*top*) Varied terrain is showcased in this Cassini mosaic of Enceladus's northern hemisphere. Cratered terrain dominates this view, with fractures attesting to a complex tectonic history. The crater at the bottom right of the image shows an area 20 km (12 mi) wide; note the domed structure in its center.
(*bottom*) Active geyserlike plumes on Enceladus were confirmed visually by Cassini's imaging system. The plumes can be seen clearly when they are backlit, as in this elegant view.

the amount of water in Saturn's environment indicates the current level of activity has lasted for at least fifteen years.

Active geyserlike plumes were finally confirmed visually by Cassini's imaging system. Multiple jets of icy material rose hundreds of kilometers above the frosted surface. Although the plumes were seen from behind, researchers deduce that they emanate from a series of canyons and ridges that border a flat region in the southern hemisphere. The terrain, extending across an area at roughly 55° south

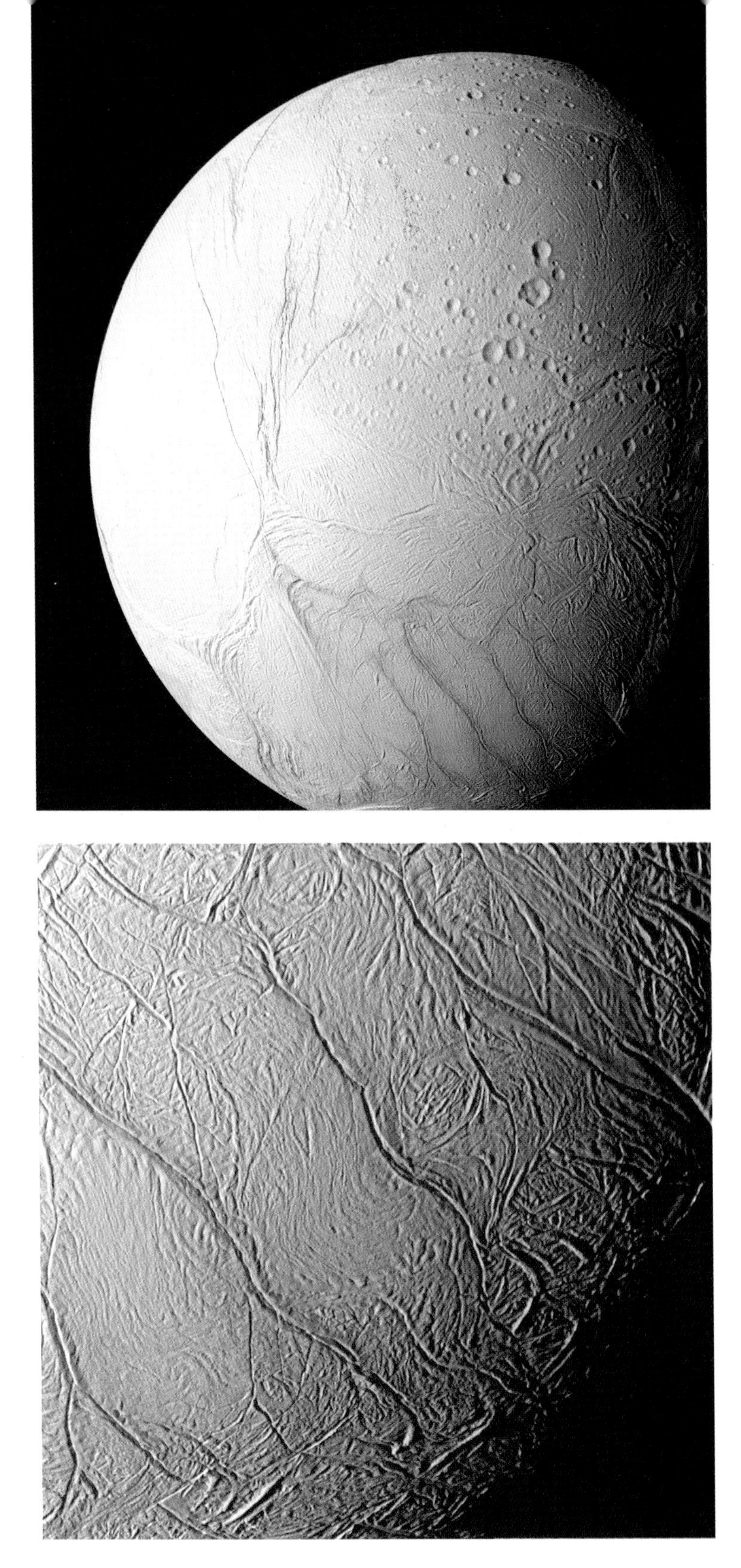

(*top*) Enceladus's tiger stripes can be seen in the lower portion of this Cassini montage. The stripes are roughly 35 km (22 mi) apart.
(*bottom*) The tiger stripes, seen here in detail, appear to be the site of the Enceladus plumes.

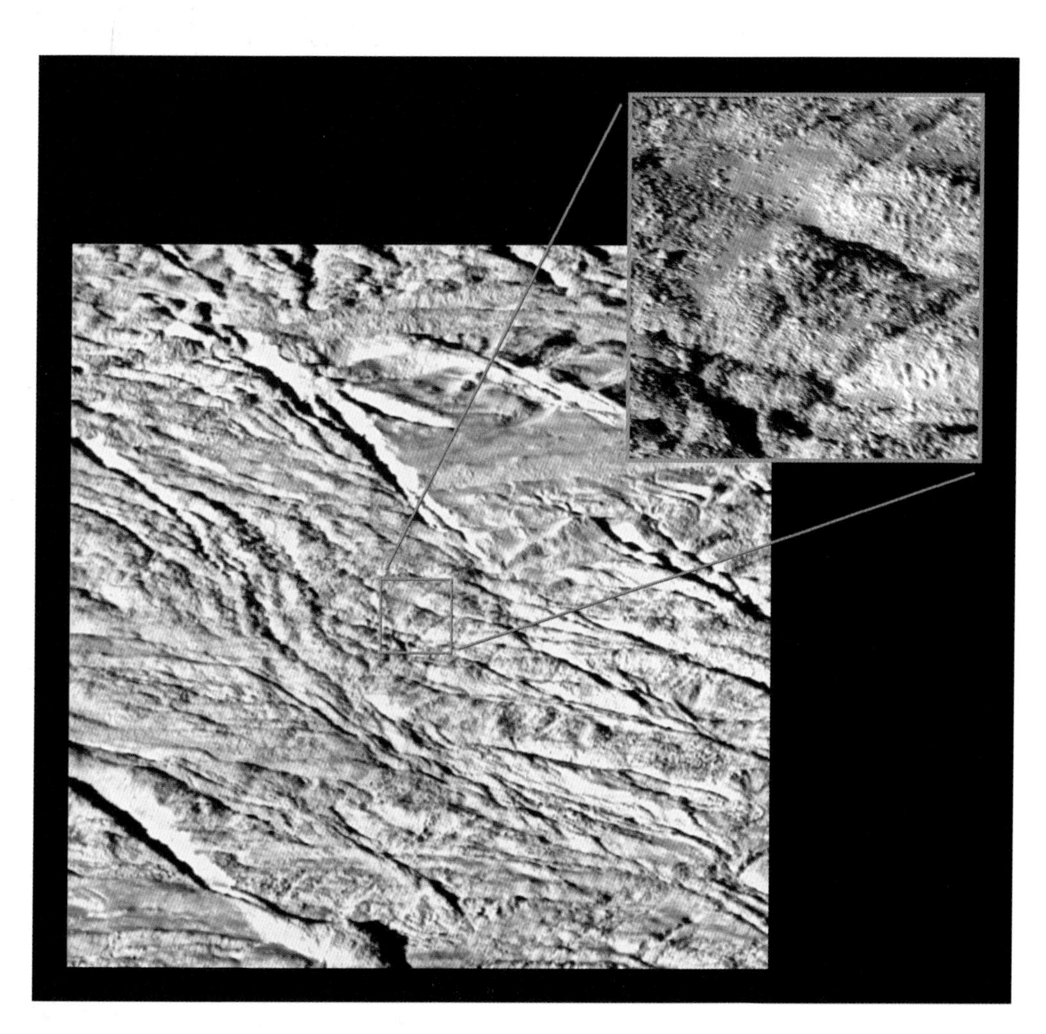

This detailed view of Enceladus shows an ice-boulder field in the vicinity of a tiger stripe. The boulders (see inset) range in size from 10 to 100 m (33–328 ft) across.

latitude, is a fresh surface scored by parallel rifts. These fractures encircle four darkened, low-lying plains called "tiger stripes." The rifts are 500 m (1,640 ft) deep, bracketed by ridges 100 m (330 ft) high. Each is about 2 km (1 mi) across, and up to 130 km (81 mi) long. Dark material that extends several kilometers to each side is apparently erupting or seeping from the rifts. The stripes are roughly 35 km (22 mi) apart.

The thermal inertia (resistance to change in temperature) of the surface is 100 times less than that of solid water ice, which suggests that the landscape is "fluffy," that is, covered in fresh ice or snow. The region plays host to dramatically elevated temperatures that reach the freezing point of water (compared to the surrounding daylight surface temperature of –201°C [–330°F]). Cassini's CIRS (composite infrared spectrometer) instrument showed that the heat is concentrated along the tiger stripes. In contrast, the heavily cratered northern hemisphere of the moon has temperatures consistent with the simple action of sunlight on a cold ice surface.

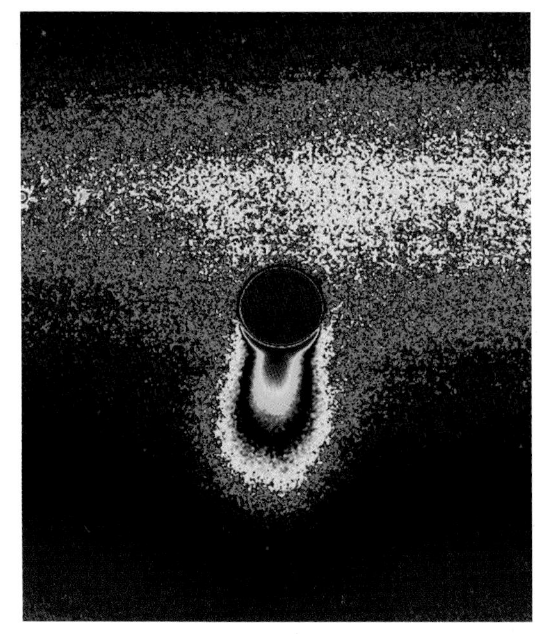

(*top*) Plumes from Enceladus are seen against the dramatic backdrop of Saturn and its rings.
(*bottom*) A plume structure is shown in this color-enhanced image of Enceladus.

There are also differences in composition between the tiger stripes and the rest of the satellite. Most of Enceladus's surface is composed of almost pure water ice, except in the region of the tiger stripes, where Cassini detected organics and CO_2 (carbon dioxide).

Indications are that the plumes eject material at a velocity of 60 m/sec (200 ft/sec). The eruptions are massive enough to resupply the E ring and infuse the entire Saturnian system with oxygen. (Sunlight breaks the plumes of water into hydrogen and oxygen, which feed into Saturn's environment.)

Temperatures in the plume itself are estimated to be at least as high as 180K (–93°C [–136°F]). This temperature is consistent with a water/ammonia content, which is one proposed mix for cryolavas at Enceladus. Oddly, no ammonia was detected in the plume. Temperatures are also consistent with a heat source roughly 660 m (2,200 ft) across, which fits well with the highest resolution images of the region.

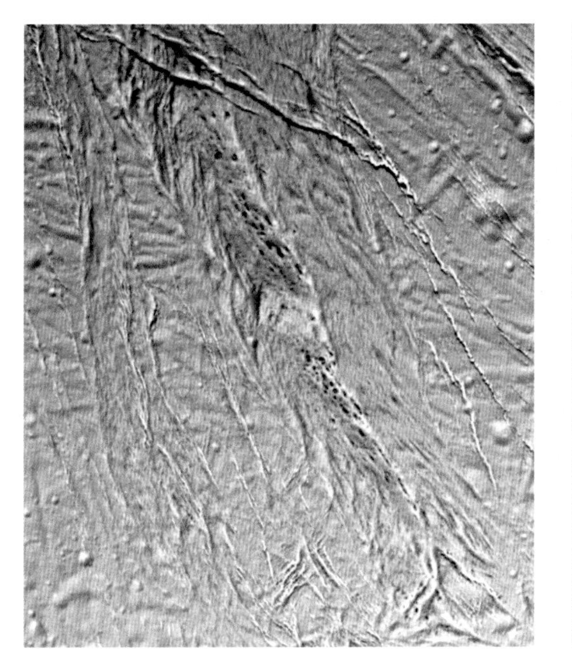

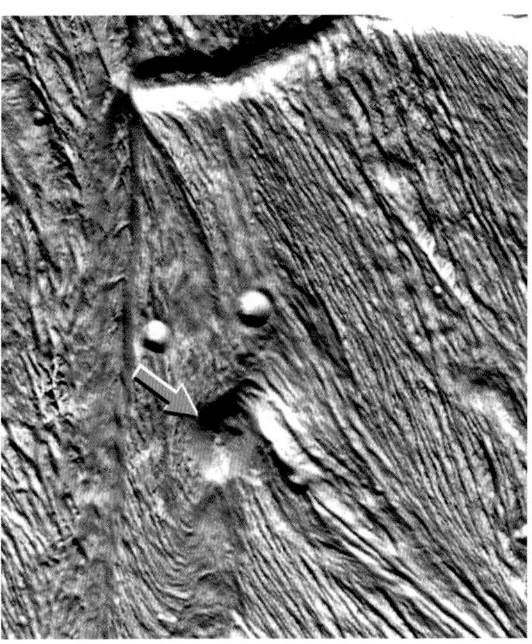

(*left*) **Dark spots among the fissures of Enceladus may be caused by internal eruptions of darker material, but evidence is not yet conclusive. The spots are about 125 to 750 m (410 to 2,460 ft) wide.**
(*right*) **This close-up view of Enceladus shows ice flows around a pitted cone that may be a cryovolcanic structure.**

What is causing cryovolcanism on the tiny worldlet? One early theory proposed water vapor blown off the surface by solar wind as water ice sublimated into gas, but the velocities seen are too great to account for such a gentle process. High temperatures within an ice surface can be generated by the solid-state greenhouse effect (as on Triton), but the Enceladus plumes seem to have too much material and force for this to be the case. Tidal heating may be the cause of volcanism on Enceladus. If Enceladus oscillates between high and low activity as it drifts in and out of various tidal stresses, the moon might have been more gravitationally stressed recently and is currently in a cool-down period. Today, Enceladus's orbit puts it in a gravitational dance with Dione, another Saturnian moon, which could contribute substantially to internal heating.

Enceladus is radiating several gigawatts of heat from within; the source of this internal heat is still unknown. The plumes of Enceladus may be linked to reservoirs of liquid water beneath its surface. As on Triton, these geyser-feeding underground lakes may exist within tens of meters of the surface. Researchers are considering two potential mechanisms for the eruptions. One is that sublimating ice builds pockets of vapor near the surface. When the pressure finally breaches the surface, a fountain of water vapor erupts. The second possibility is that underground reservoirs of boiling liquid, perhaps a mix of water and ammonia or water and methane, erupts through the linear vents in the tiger stripes. Cassini detected a high ice to gas ratio in the plumes, which implies that ice is not condensing out of pure vapor, as

ENCELADUS IN MYTHOLOGY

In Greek mythology Enceladus was one of the race of superhuman giants who warred with the gods of Olympus. These giants were born from the blood of Uranus, the son and husband of Gaia (earth), who was castrated by his own son Chronus. Enceladus and the other giants were defeated by the gods of Olympus, and Enceladus was imprisoned inside Mount Etna. As the giant struggled to free himself, the earth shook; his voice rumbled within the volcano; his searing breath tossed molten rock into the sky. In light of Cassini's discoveries at Enceladus's namesake moon, the ancient legend seems particularly apropos.

the first option posits. This second theory favors a mix of water with something that will lower its boiling point. It is possible that ammonia reserves combine with water underground, creating a boiling brew, but the ammonia must not be venting through the surface with the water, because Cassini detected none in the atmosphere.

Enceladus's unique volcanism has far-reaching implications. It opens possibilities for finding active geology on bodies once thought to be too small for such processes. It also multiplies the potential sites for exobiology. Subsurface water on Enceladus may have been stable for long periods, perhaps millions or billions of years. This fact, coupled with evidence for organic material at the eruptive sites, makes Saturn's moon an intriguing target for astrobiologists. Tiny Enceladus joins the ranks of Mars and Europa as a site where future explorers may search for life among alien volcanoes.

INTO THE GREAT BEYOND

What will we find in other planetary systems? Is volcanism likely on worlds orbiting Proxima Centauri or Betelgeuse? Wherever we find terrestrial planets, we will almost certainly find evidence for past or present volcanism similar to that seen on the terrestrial planets of our own system. We know that a wide range of volcanism exists on satellites of giant planets, and these probably exist in other solar systems. But because of exotic chemistries and alternative star arrangements, volcanoes there may be driven by new and unimagined mechanisms. Our sun is fairly rare in that it is a single star. The majority of star systems contain pairs

or multiple members, dancing a complicated cotillion of orbits around each other. Any nearby planets will probably be subject to eccentric orbits and variable tidal friction. Perhaps some planets have eruption cycles concurrent with their orbital dynamics.

Even more varied eruptions must take place on icy worlds yet unseen. Moons of gas giants—which seem to abound in nearby stellar systems—will be subject to forces similar to those on Io, Enceladus, and Triton. Surely other alien volcanoes await us when we venture from our solar system into the great beyond.

Volcanic activity and climate change are fired by the elliptical orbit of a planet around a double star system. Stars frequently appear as twins or triplets, and many planets have been detected in orbit around nearby stars. The artist's view envisioned here may be fairly common in the galaxy.

8 SPACE, VOLCANOES, AND CULTURE

J. R. R. Tolkein gave us Mount Doom, with its thundering clouds and exploding fireballs. Jules Verne took us down the very throat of Iceland's Mount Snaefells to get to another world. Something in the nature of volcanoes touches a visceral level in us all, something primordial and elemental. Strewn across the globe in widespread patterns, volcanoes are an integral part of the lives of many peoples. In every inhabited continent people have learned to coexist with thundering mountains, finding their own explanations and ways of coping within the shadow of earth's raw power. These looming shadows have been celebrated throughout recorded history, in writing, art, and song that reach across the cultures and religions of the world.

The Roman god Vulcan (Hephaistos in Greek mythology) had a forge with a chimney that led to the summit of the volcano known as Vulcano. Homer says the giant Polyphemus lived in a cave under Sicily's famous Mount Etna. When Odysseus and his men escaped from Polyphemus, he "threw rocks at them." Some scholars interpret this is as a reference to volcanic activity.

Half a world away, an important figure in the ancient Hawaiian pantheon was Madame Pele, goddess of volcanoes. This beautiful, tempestuous, and unpredictable goddess vented her anger through calderas and lava fountains. Like many ancient deities, Pele was at once feared and revered. When she lost her temper, she was known to trigger earthquakes by stamping her feet and to set off volcanic eruptions by digging with her magic stick, called the Pa'oe. One legend portrays Pele quarreling with her older sister,

(*opposite*) Madame Pele, the Hawaiian volcano goddess, is imagined here rising from the cone of a volcano.

Namakaokaha'i. The violent argument led to the creation of the series of volcanoes that form the Hawaiian chain.

Native American civilizations had legends—and perhaps sacrifices—related to volcanoes. At the 6,705.6-m (22,000-ft) summit of an Andean volcano, three mummified children were discovered in 1999. The five-hundred-year-old mummies were well preserved in the cold, dry environment. Farther to the north, the Aztecs named twin volcanoes near their capital Popocatepetl (the Smoking Mountain) and Ixtaccihuatl (the White Lady). They were thought to be tumultuous lovers.

Volcanoes figure prevalently in the mythology of more northerly Native American legends. The Klamath tribe of the northwestern United States tells the legend of the fire god Llao, who lived inside Mount Mazama, the volcano whose collapse created Crater Lake in Oregon. Skell was a mighty snow spirit whose dominion included Mount Shasta. A great war broke out between the forces of the two spirits. Skell was killed and his heart taken to the shores of Crater Lake for a celebration. The followers of Skell stole the heart back. When Skell's heart was restored to him, he resumed the war and defeated Llao. Llao's body sank into Mount Mazama, opening a crater in the mountaintop. His body was torn into pieces, but his head was preserved at the lake's center, where it is known today as Wizard Island. Other native legends tell of battles between volcanic brothers Wy'east (Mount Hood) and Pahto (Mount Adams), who fought over the lovely La-wa-la-clough (smoking mountain), more commonly known today as Mount Saint Helens.

Some scholars suggest that ancient references to volcanoes crop up in biblical accounts, perhaps including Moses' smoking Mount Sinai and even the biblical plagues of Egypt. These scholars point to historic eruptions that have colored waterways with a sluice of reddish clay or mud, turning rivers to "blood." The Mosaic account describes darkness, followed by a fiery hail, conjuring up thoughts of Pliny's description of the Vesuvius eruption in AD 79. These horrors were followed by a plague of boils, which some tie to the same kinds of skin afflictions seen in medical centers at the time of the Mount Saint Helens catastrophe. This plague was followed by the widespread death of cattle and fish. The toxic Nile, experts reason, would have given rise to dead creatures of the waterways, followed by flies that would attract frogs and other creatures. A link between these biblical events and the eruption of Santorini around 1600 to 1650 BC has been suggested, but dates for both the eruption and the biblical story are uncertain.

Perhaps the most intriguing tie-in between volcanic activity and ancient writings is that concerning Plato's Atlantis. Modern scholars have hotly debated the historic-

ity and possible location of this legendary civilization. Plato wrote of Atlantis in 400 BC. Pliny the Elder added an account in the first century AD, although his seems to have been based primarily on Plato's work. The heart of this advanced civilization was a series of islands arranged in nine concentric rings connected by bridges. Canals allowed ships to pass from one circular waterway to the next. The metropolis on the central island was made of red and black stone quarried from local sources; roofs in Atlantis were copper. The islands' technology was advanced beyond most of the surrounding civilizations. A racecourse ringed the island, and two central temples glistened with gold, silver, and ivory. Public baths, as well as baths in homes, were heated by natural hot springs. The Atlanteans were also renowned seafarers. Plato tells us that the great civilization was lost when a series of earthquakes and floods swamped the island and it "sank into the sea."

Plato's description of the legendary Atlantis has been linked by some to the island of Thera, now known as Santorini, in the Aegean Sea. The economic center of a

Steep cliff walls (*bottom right*) are all that is left of the caldera of Santorini (ancient Thera). The island volcano, which was destroyed in a cataclysmic explosion thirty-five hundred years ago, may have given rise to the legend of Atlantis. The island in the background is Nea Kameni, an active volcano.

VOLCANOES IN POETRY

Volcanoes have influenced many art forms throughout history, from paintings to poetry to short stories to dance. In this poem, the space artist and explorer Joel Hagan reflects on the awesome power of a volcano's breath.

The Long Breath

Plutonic pulse of molten stone
Million-year heartbeat
Breathe
And mountains are born and torn asunder
Mend and rend the agile crust
Plunge to magma and breathe again

We are not your skittering dust of life
But we take you into our realm
Teach fragile bone the craft of stone
Gather swimmers to our pelagic belly
Gifting them with eons

You hoppers and spinners
Flitter and prance imagined journeys
Yet have not moved

We are the only place
Born in spacious beauty at each journey's end
Joined with kin fresh from the void

No death
Transformation, journeys,
And the long breath
Of place

grand civilization has been discovered on Santorini, which is the island remnant of a volcano that exploded between 1650 and 1600 BC. The buildings are made of red and black cut stone; the roofs are copper; and the city was apparently wealthy. The Minoan civilization that held court there was well known in ancient times for its maritime network and its trade routes, which linked all of the Mediterranean. Egyptian writings give us an idea of how influential this seafaring nation was. The Egyptians referred to all foreign people in dismissive or derogatory tones—except for the Minoans, whom they called the Sea People.

The city on ancient Thera had public baths heated geothermally, and even some private homes had indoor plumbing. The descriptions of Atlantis are hauntingly familiar to those who study ancient Thera and the Minoans. The sinking of Atlantis may well be a description of the destruction of Thera. The volcanic island succumbed to an explosive event far more powerful than that of Krakatau. The blast sent tidal waves into Turkey and moved one-ton

In this artist's rendering, a vast lake of molten lava spreads before the plumes of Io's volcano Loki. Within the lake stands a sulfur dioxide island the size of Rhode Island.

blocks of stone on Crete, 97 km (60 mi) to the south. Thera may, in fact, have inspired the legend of Atlantis.

Volcanoes are also found in more modern writing. Emily Dickinson's "The reticent volcano keeps" likens human emotion to the spectacular geological events found in volcanoes. Wallace Stevens gave us "Postcards from the Volcano" and Susan Sontag, *The Volcano Lover*. *Under the Volcano,* by the British novelist Malcolm Lowry, uses the imagery of brooding Mexican volcanoes to illuminate a day in the tragic life of an alcoholic. In one of Joanna Cole's magic school bus adventures, the bus and its passengers escape from the center of the earth with the aid of a helpful volcano's belch.

Volcanoes have also inspired dance and song. Hawaiian hula dances celebrate Pele and her handiwork. The Tolai people of Rabaul in Papua New Guinea have blended choral and stringed music as they sing of their relationship to the great volcanoes that engender rich farmlands while taking away life in waves of ash and fire. Jimmy Buffet penned

PAYING TRIBUTE

Throughout history, people have bestowed offerings on the flanks of volcanoes. At times, these gifts were given in worshipful adoration; at others, they were presented in hopes of taming lava flows or stemming eruptions.

Hawaiians regularly sang songs to the volcano goddess, Madame Pele; left food and tobacco for her; and sometimes even poured gin into fuming calderas. In preconquest Mexico, Aztecs made cornmeal volcanoes each October. The cone-shaped cakes were crowned by a head. After three weeks, the cakes were decapitated and eaten. The ancient peoples of Nicaragua sacrificed an infant every twenty-five years to appease the volcano Cosigüina.

Even today, people around the world pay tribute to erupting mountains. Each October in eastern Java, thousands gather to celebrate the Kesodo festival. Priests lead the long procession three times around the rim of Mount Bromo while the people throw money, flowers, and even live chickens into the caldera. In Japan, tourists and residents gather together annually on the flanks of the great Fujiyama in meditation and contemplation.

Regardless of the scientific knowledge we have gained about volcanoes—terrestrial or alien—these smoldering mountains still have the power to touch something deep within the human psyche. We are, in the truest sense, awed by these growling giants. Let us hope we always will be.

the words "I don't know where I'm a gonna go when the volcano blows."

Now that we have discovered erupting summits beyond our home world, the myths of the ancients have cast their long shadows beyond our home planet. Hawaii's Madame Pele is engraved upon our new maps of Io, where her cosmic namesake erupts sulfur dioxide into a black sky. The Norse god Loki has taken his place there, too, bellowing sulfurous plumes and draping snowy blankets across the Ionian landscape. Io sports many such monikers from the mythologies of fire and thunder around the world. Prometheus, the Greek god who gave fire to humankind, makes landfall as a massive volcano on Io's Jupiter-facing side. Shamash and Marduk add their names from Mesopotamian mythology, as do Surya and Vivasant from Hindu cultures and Tupan and Monan from Brazilian nature myths. As humankind travels to the distant worlds around us, we give the alien volcanoes names from our world mythology and make them our own.

INDEX

ABOUT THE AUTHORS

Dr. Rosaly M. C. Lopes is lead scientist for geophysics and planetary geosciences at NASA Jet Propulsion Laboratory and investigation scientist for the Titan Radar Mapper on the Cassini spacecraft. Lopes started working in volcanology while a graduate student in London. She has visited many of earth's most active volcanoes and has studied volcanism of other planets and satellites in our solar system. She worked for several years on NASA's Galileo mission studying Jupiter's moon Io, where she discovered seventy-one new active volcanoes—which earned her a mention in the 2006 *Guinness Book of World Records*. She currently works on the Cassini mission studying ice volcanism on Titan, Saturn's largest moon. She has written numerous academic papers, as well as articles for popular publications. She is the author of *The Volcano Adventure Guide* (Cambridge University Press, 2006), the first travel guidebook to volcanoes, and is the editor of *Volcanic Worlds: Exploring the Solar System Volcanoes* (Praxis-Springer, 2004) and *Io after Galileo* (Praxis-Springer, 2007). She is a member of several scientific societies and a Fellow of the American Association for the Advancement of Science and the Explorers Club. She frequently lectures to the public and is featured on TV documentaries. In 2005 she was awarded the Carl Sagan Medal by the Division for Planetary Sciences of the American Astronomical Society for her efforts in communicating science to the public. For more information about Lopes go to www.volcano adventures.com.

Michael W. Carroll has loved volcanoes since his parents introduced him to one in Hawaii when he was eleven. Carroll has been painting volcanoes, space, and other things for more than two decades. He has done work for NASA and the Jet Propulsion Laboratory, and his art has appeared in several hundred magazines throughout the world, including *Time, National Geographic, Sky and Telescope, Asimov's, Smithsonian, Astronomy, Ciel et Espace, and Astronomy Now* (UK). Carroll is a Fellow and founding member of the International Association for the Astronomical Arts and a member of the NASA Arts Program. One of his original paintings flew aboard Mir; another is resting at the bottom of the Atlantic aboard Russia's ill-fated Mars 96 spacecraft. He has painted murals in numerous science museums and planetariums across the United States. He has written science articles for many magazines and has coauthored, with his wife, Caroline, a dozen science books for children and adults. His work can be seen at www.stock-space-images .com and www.spacedinoart.com.

CREDITS

All art is by Michael W. Carroll, except Hieronymus Bosch, c. 1500, *The Garden of Earthly Delights,* courtesy Museo de Prado, Madrid, 2; Paul Kane, 1874, *Mount Saint Helens Eruption,* courtesy Stark Museum, Orange, Texas, 21; and Chesley Bonestell, *Jupiter Volcano,* in *The Solar System* (Chicago: Childrens Press, 1961), 39. Photographs by Michael W. Carroll are on pages 3 (top), 5, 7, 17, 27 (left), 28 (bottom), 29, 31 (top and bottom), 32, 92 (bottom), and 113 (top).

All other art appears courtesy of the following:

Elsa Abbott, JPL: 34 (bottom)
Katharine Cashman: 8
Chris Drake: 10 (top)
ESA: ExoMars lander: 53 (right)
ESA/NASA/JPL/University of Arizona: 123 (top) (PIA08114)
Charles Frankel: 12
John E. Guest: 18
William K. Hartmann: 19
Rosaly M. C. Lopes: 34 (top)
Malin Space Science Systems: 48
Lesa Moore, Macquarie University: 84 (top)
NASA/Johns Hopkins University Applied Physics Laboratory/Southwest Research Institute: 87 (PIA09256)
NASA/JPL: 47
NASA/JPL/Arizona State, THEMIS images: 44 (PIA08491), 45 (left) (PIA06827), 45 (right)
NASA/JPL-Caltech: 24 (PIA00404), 38 (top and bottom), 40, 41 (top left) (PIA00993), 42 (PIA01457), 52 (top) (PIA00289), 53 (left) (PIA08553, Mars Science Lab), 55 (PIA00245), 56 (PIA00200), 57 (PIA00202), 58 (PIA00101), 60 (PIA00215), 64 (PIA00307), 65 (top) (PIA00201), 65 (bottom) (PIA00089), 66 (PIA00103), 78 (PIA03102), 86 (top) (PIA01299), 86 (bottom) (PIA01074), 88 (left) (PIA02254, PIA01362), 88 (right) (PIA01368), 89 (top) (PIA00379), 89 (bottom) (PIA00585), 91 (top) (PIA02526), 91 (bottom) (PIA02565), 92 (top) (PIA02557), 93 (PIA02550), 94 (PIA01667), 97 (top) (PIA01254), 100 (top) (PIA01656), 100 (bottom) (PIA00056), 102 (PIA01127, closeup PIA01181), 103 (PIA00586, PIA01211), 106 (top) (PIA01647), 106 (bottom) (PIA00590), 107 (top) (PIA00849), 109, 110 (PIA01640), 115 (PIA01614), 120, 122 (top and bottom), 123 (bottom) (PIA06993), 125 (PIA00317), 126 (top) (PIA01538), 126 (bottom) (PIA00061), 127, 128 (PIA00060), 130 (PIA08163), 131 (bottom) (PIA01950), 132 (PIA06248), 134 (bottom) (PIA07758)
NASA/JPL-Caltech/Cornell: 49 (bottom) (PIA06102)
NASA/JPL-Caltech/Cornell/NMMNH: 49 (top) (PIA08440)
NASA/JPL/Malin Space Science Systems: 52 (left) (PIA04676)
NASA/JPL/Northwestern University: 79 (right) (PIA02434), 79 (left) (PIA02943)
NASA/JPL/Space Science Institute: 116 (PIA00038), 133 (top) (PIA06207), 134 (top) (PIA07800), 134 (bottom) (PIA06247), 135 (PIA06252), 136 (top) (PIA08199), 136 (bottom) (PIA08226), 137 (left) (PIA06188)
NASA/JPL/University of Arizona: 43 (left and right), 96 (PIA02599)
NASA/JPL/University of Arizona/University of Colorado: 111 (PIA03878)
NASA/JPL/USGS: 137 (right)
NASA/JPL/USGS/NRL/BMDO: 70
NASA/JSC: 71 (top and bottom), 72, 73, 74, 75
NASA/MER team: 41 (top right)
Scott Rowland: 3 (bottom), 10 (bottom), 28 (top), 35
Florian Schwandner: 143
Joe Smyth: 27 (right)
Ted Stryk and the Russian Academy of Sciences: 54
USGS: 13, 41 (bottom)
USGS, Cascades Volcano Observatory, Lyn Topinka: 23 (top and bottom)
USGS, M. Doukas: 22
USGS, Mark Robinson and Paul Lucey: 80 (left and right)
Ralph B. White: 113 (bottom)

Sources for quotations

5 "Then came a man running," Njal's Saga, c. AD 1280, author unknown. *Njal's Saga,* ed. and trans. Robin Cook (London: Penguin, 2001), 101–2.

7 "It rose into the sky," Pliny the Younger, *The Letters of the Younger Pliny,* trans. Betty Radice (Baltimore: Penguin, 1963).

9 "A volcano is an immense cannon," Georges-Louis Leclerc de Buffon, *Histoire Naturelle* (1749).

15 "It is just as if we were to refit the torn pieces," Alfred Wegener, *The Origin of Continents and Oceans* (1915).

37 "These disturbances may be called 'volcanic,'" Willy Ley, *Conquest of Space* (New York: Viking, 1949).

85 "Philosophy is written in this grand book," Galileo Galilei, *Il Saggiatore* [1623].

121 "methane monsoons," Arthur C. Clarke, *Imperial Earth* (New York: Harcourt Brace Jovanovich, 1976).

144 "The Long Breath," reprinted by permission of the author.